THE DEFINITIVE ILLUSTRATED GUIDE TO
THE ELEMENTS

JACK CHALLONER studied physics in London and trained as a science and maths teacher. He then worked at London's Science Museum, in the education department and in their flagship interactive gallery, Launch Pad. Jack left the museum in 1991 to write science and technology books, and to write and perform science shows in museums, libraries and schools. He's been doing all that ever since, with nearly 40 books to his name including The Cell, Genius: Great Inventors and their Inventions and Rocks & Minerals. He also works as a consultant on other people's science books, and sometimes in television, helping to develop new ideas or work up existing ones. He is also a musician and singer, writing, producing and performing music. He lives in Bristol, England.

This edition first published in 2016 by Andre Deutsch
An imprint of the Carlton Publishing Group
20 Mortimer Street
London W1T 3JW

Hardback edition first published in 2012

A CIP catalogue record for this book is available from the British Library.

ISBN 978 0 233 00488 4

Printed in China

10 9 8 7 6 5 4 3 2 1

THE DEFINITIVE ILLUSTRATED GUIDE TO
THE ELEMENTS

INTRODUCING THE BUILDING BLOCKS OF OUR UNIVERSE

JACK CHALLONER

ANDRE
DEUTSCH

Contents

1 H Hydrogen								
3 Li Lithium	4 Be Beryllium							
11 Na Sodium	12 Mg Magnesium							
19 K Potassium	20 Ca Calcium	21 Sc Scandium	22 Ti Titanium	23 V Vanadium	24 Cr Chromium	25 Mn Manganese	26 Fe Iron	27 Co Cobalt
37 Rb Rubidium	38 Sr Strontium	39 Y Yttrium	40 Zr Zirconium	41 Nb Niobium	42 Mo Molybdenum	43 Tc Technetium	44 Ru Ruthenium	45 Rh Rhodium
55 Cs Caesium	56 Ba Barium		72 Hf Hafnium	73 Ta Tantalum	74 W Tungsten	75 Re Rhenium	76 Os Osmium	77 Ir Iridium
87 Fr Francium	88 Ra Radium		104 Rf Rutherfordium	105 Db Dubnium	106 Sg Seaborgium	107 Bh Bohrium	108 Hs Hassium	109 Mt Meitnerium

57 La Lanthanum	58 Ce Cerium	59 Pr Praseodymium	60 Nd Neodymium	61 Pm Promethium	62 Sm Samarium
89 Ac Actinium	90 Th Thorium	91 Pa Protactinium	92 U Uranium	93 Np Neptunium	94 Pu Plutonium

2
He
Helium

Alkali Metals
Alkaline Earth Metals
Transition Metals
Post-Transition Metals
Metalloids
Other Non-Metals
Halogens
Noble Gases
Lanthanoids
Actinoids
Transuranium Elements

5 **B** Boron	6 **C** Carbon	7 **N** Nitrogen	8 **O** Oxygen	9 **F** Fluorine	10 **Ne** Neon
13 **Al** Aliminium	14 **Si** Silicon	15 **P** Phosphorus	16 **S** Sulfur	17 **Cl** Chlorine	18 **Ar** Argon

28 **Ni** Nickel	29 **Cu** Copper	30 **Zn** Zinc	31 **Ga** Gallium	32 **Ge** Germanium	33 **As** Arsenic	34 **Se** Selenium	35 **Br** Bromine	36 **Kr** Krypton
46 **Pd** Palladium	47 **Ag** Silver	48 **Cd** Cadmium	49 **In** Indium	50 **Sn** Tin	51 **Sb** Antimony	52 **Te** Tellurium	53 **I** Iodine	54 **Xe** Xenon
78 **Pt** Platinum	79 **Au** Gold	80 **Hg** Mercury	81 **Tl** Thallium	82 **Pb** Lead	83 **Bi** Bismuth	84 **Po** Polonium	85 **At** Astatine	86 **Rn** Radon
110 **Ds** Darmstadtium	111 **Rg** Roentgenium	112 **Cn** Copernicium	113 **Uut** Ununtrium	114 **Fl** Flerovium	115 **Uup** Ununpentium	116 **Lv** Livermorium	117 **Uus** Ununseptium	118 **Uuo** Ununoctium

63 **Eu** Europium	64 **Gd** Gadolinium	65 **Tb** Terbium	66 **Dy** Dysprosium	67 **Ho** Holmium	68 **Er** Erbium	69 **Tm** Thulium	70 **Yb** Ytterbium	71 **Lu** Lutetium
95 **Am** Americium	96 **Cm** Curium	97 **Bk** Berkelium	98 **Cf** Californium	99 **Es** Einsteinium	100 **Fm** Fermium	101 **Md** Mendelevium	102 **No** Nobelium	103 **Lr** Lawrencium

Introduction

"Modern physics and chemistry have reduced the complexity of the sensible world to an astonishing simplicity." – *Carl Sagan*

Elements, compounds and mixtures

Most familiar substances are mixtures or compounds. Wood, steel, air, salt, concrete, skin, water, plastics, glass, wax – these are all mixtures or compounds, made up of more than one element.

We do encounter elements in our everyday life, albeit not completely pure. Gold and silver are good examples; and even in the purest sample of gold ever produced, one in every million atoms was an atom of an element other than gold. Copper (pipes), iron (railings), aluminium (foil) and carbon (as diamond) are further examples of elements we encounter in their fairly pure state. Some other elements are familiar simply because they are so important or commonplace. Oxygen, nitrogen, chlorine, calcium, sodium, lead – these are all examples of such elements.

This book will explore the properties of all the elements. The properties of an element include its chemical behaviours – in other words, how its atoms interact with atoms of other elements. So for each element, we will also look at some important compounds and mixtures that contain it.

Please read!

Sometimes, it makes little practical difference whether you read a book's introduction or not. But that is not the case here. This introduction contains crucial information that will enable you to understand the organization of this book and the information it contains. It will also help you appreciate the complex beauty of the world – and how all of it can be explained by the interactions between only three types of particle: protons, neutrons and electrons. For it is a mind-boggling truth that from the core of our planet to the distant stars, all matter – be it solid, liquid, gas or plasma – is made of different combinations of just these three particles.

Protons, neutrons and electrons

An atom has a diameter in the order of one ten-millionth of a millimetre (0.0000001 mm, 0.000000004 inches). An atom's mass is concentrated in a heavy central part, the nucleus, made of protons and neutrons.

The much lighter electrons surround the nucleus. Everything around you is made of only about 90 different types of atom: 90 different arrangements of protons, neutrons and electrons. These different *types* of atom are the **elements**.

Protons carry positive electric charge; electrons carry a corresponding amount of negative electric charge. Scale them up in your imagination, so that they are little electrically charged balls you can hold in your hand, and you would feel them pulling towards each other because of their mutual electrostatic attraction. Neutrons, as their name suggests, are neutral: they carry no electric charge. Hold a scaled-up one of these in your hand, and you will see that it is not attracted towards the proton or the electron.

Building atoms

With these imaginary, scaled-up particles, we can start building atoms of the first few elements – beginning with the simplest and lightest element, hydrogen.

Proton, p⁺

Neutron, n

Electron, e⁻

Above: Illustration of a proton (red), neutron (blue) and electron. The mass of a proton is the same as that of a neutron, more than 1,800 times that of an electron.

For the nucleus of your hydrogen atom, you just need a single, naked proton. To that, you will need to add your electron – by definition, an atom has equal numbers of protons and electrons, so that it has no charge overall. Hold the electron at some distance from the proton and the two particles will attract, as before. The force of attraction means that the electron has potential energy. Let go of the electron and it will "fall" towards the proton losing potential energy. You will notice that it stops short of crashing into the proton, and settles instead into an orbit around it. It is now in its lowest energy state.

Strange behaviours

You have just built a hydrogen atom – albeit an imaginary one. There are a few strange things to notice, for the world of tiny particles is dominated by the weird laws of **quantum physics**. For example, as your electron fell towards your proton, you will have noticed that it did so in distinct jumps, rather than one smooth movement. For some reason that is built into the very fabric of the Universe, the electron is only "allowed" certain energies. The amount of energy the electron loses in each jump – the difference in energy between any two levels – is called a **quantum**. The lowest level of potential energy, which corresponds to the electron's closest approach, is often written **n=1**.

Every quantum of energy lost by an electron creates a burst of visible light or ultraviolet radiation, called a **photon**. Any two photons differ only in the amount of energy they possess. A

Above: Illustration showing the distinct electron energy levels around a hydrogen nuleus and around a beryllium nucleus (not to scale).

photon of blue light has more energy than a photon of red light, and a photon of ultraviolet radiation has more energy than a photon of blue light. If you now knock your electron back up a few levels, watch it produce photons as it falls back. Some of the photons will be visible light, others will be invisible ultraviolet ones.

Each element has a characteristic set of energy levels, since the exact levels are determined by the number of protons in the nucleus. And so, each

Left: Discrete (separate and well-defined) lines in the visible part of the spectrum, produced by excited hydrogen atoms.

element produces a characteristic set of photons of particular frequencies, which can be examined using a prism to separate the different frequencies into a spectrum consisting of bright lines on a dark background.

As a consequence, elements can be identified by the colours of the light they give out when their electrons are given extra energy (excited) then allowed to settle down again. You can excite an electron with heat, electricity or by shining ultraviolet radiation on to it. Metal atoms will produce characteristic coloured light in the heat of a flame, for example – see page 23 for pictures of flame tests; and this process is responsible for the colours of fireworks, as electrons in metal atoms are repeatedly excited by the heat of combustion and then fall down to lower energies again. And in energy-saving fluorescent lamps, ultraviolet radiation excites electrons in atoms in the glass tube's inner coating, producing red, green and blue photons that, when entering the eye together, give the illusion of white light.

Fuzzy orbitals

You will have noticed another strange behaviour in your imaginary atom. Instead of being a well-defined particle, your electron appears as a fuzzy sphere surrounding the nucleus, called an **orbital**. The quantum world is an unfamiliar, probabilistic place, in which objects can be in more than one place at the same time and exist as spread-out waves as well as distinct particles. And so as well as being a well-defined particle, your electron is also a three-dimensional stationary wave of probability. The chemical properties of elements are determined mostly by the arrangement of electrons in orbitals around the nucleus.

Atomic number

Now move the electron away, leaving the naked proton again. To make the next element, with atomic number 2, you will have to add another proton to your

Above: Illustration of an orbital, the region in which electrons can exist – as both a point particle and a spread-out wave.

nucleus. But all protons carry positive charge, so they strongly repel each other. What is worse, the closer they get, the more strongly they push apart. Fortunately, there is a solution. Put the second proton down for a moment and try adding a neutron instead. There is nothing stopping you this time, because the neutron has no electric charge.

As you bring the neutron very close, you suddenly notice an incredibly strong force of attraction, pulling the neutron and proton together. This is the **strong nuclear force** – it is so strong that you will now have trouble pulling the proton and neutron apart. It only operates over an exceedingly short range. You now have a nucleus consisting of one proton (1p) and one neutron (1n). This is still hydrogen, since elements are defined by the number of protons in the nucleus – the **atomic number.** But this is a slightly different version of hydrogen, called hydrogen-2. The two versions are **isotopes** of hydrogen, and if you add another neutron, you will make another isotope, hydrogen-3.

The strong nuclear force works with protons, too (but not electrons). If you can manage to push your other proton very close to your nucleus, the attractive strong nuclear force will overcome the repulsive force. The proton sticks after all, and your hydrogen-3 nucleus has become a nucleus of helium-4, with two protons and two neutrons (2p, 2n). This process of building heavier nuclei from lighter ones is called **nuclear fusion**.

Building elements
Protons and neutrons were forced together in this

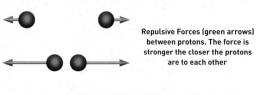

Repulsive Forces (green arrows) between protons. The force is stronger the closer the protons are to each other

Attractive Force (orange line) is the strong nuclear force between proton and neutron

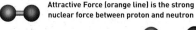

Strong nuclear force overcomes the repulsion between two protons

 hydrogen-2   hydrogen-3   helium-4

Above: Protons repel each other, and that repulsion increases the closer they are. But, at very small distances, the strong nuclear force holds protons and neutrons together, and can overcome that repulsion, building nuclei.

way in the intense heat and pressure in the first few minutes of the Universe, building elements up to beryllium-8, which has 4 protons and 4 neutrons. All the other elements have been produced since then, by nuclear fusion inside stars. For example, three helium-4 nuclei (2p, 2n) can fuse together to make a nucleus of carbon-12 (6p, 6n); add another helium-4 nucleus and you make oxygen-16 (8p, 8n), and so on. Various combinations are possible, and during its lifetime a typical star will produce all the elements up to iron, which has atomic number 26, using only hydrogen and helium as starting ingredients. Elements with higher atomic numbers can only be produced in supernovas – stars exploding at the end of their life cycle. So everything around you – and including you – is made of atoms that were built in the first few minutes of the Universe or inside stars and supernovas.

Electron shells
To the helium-4 nucleus you made you will need two electrons if you want it to become a helium atom. Drop them in towards your new nucleus and you will find they both occupy the same spherical orbital around the nucleus – an **s-orbital** (the "s" has nothing to do with the word "spherical"). The two electrons are both at the same energy level, n=1, so this particular orbital is labelled 1s. Hydrogen has an **electron configuration** of $1s^1$; helium's is $1s^2$. As you build heavier elements, with more electrons, the outermost electrons will be further and further from the nucleus, as the innermost slots become filled up.

An orbital can hold up to two electrons, so when it comes to the third element, lithium, a new orbital is needed. This second orbital is another spherical s-orbital, and it is at the next energy level, n=2, so

helium-4 nucleus

beryllium-8 nucleus (two helium-4 nuclei)

carbon-12 nucleus (three helium-4 nuclei)

oxygen-16 nucleus (four helium-4 nuclei)

Above: Building larger nuclei. Inside stars some of the most common elements are formed by the fusion of helium-4 nuclei. Shown here are beryllium-8, carbon-12 and oxygen-16

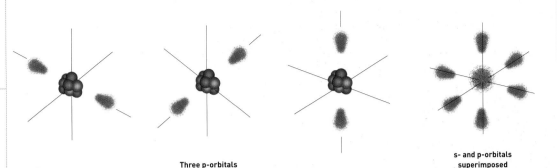

Three p-orbitals

s- and p-orbitals
superimposed

it is labelled 2s. The electron configuration of lithium is $1s^2 2s^1$. If you look at the periodic table on the contents page, on page 5, you will see that lithium is in the second row, or period. The rows of the periodic table correspond to the energy levels in which you find an atom's outermost electrons. So hydrogen and helium are in the first period because their electrons are at n=1. The Period 2 elements, from lithium to neon, have their outermost electrons at energy level n=2.

Electrons that share the same energy level around the nucleus of an atom are said to be in the same **shell**. The electrons of hydrogen and helium can fit within the first shell (energy level n=1). At the second energy level – in the second shell – there is more space for electrons. A new type of orbital, the dumbbell-shaped **p-orbital**, makes its first appearance. Like an s-orbital, a p-orbital can hold up to two electrons. There are three p-orbitals, giving space for six electrons. So the second shell contains a total of eight electrons: two in s- and six in p-orbitals. Neon, at the end of Period 2, has the electron configuration $1s^2 2s^2 2p^6$, and has a filled outer shell; the element neon has an atomic number of 10.

The third shell also has one s- and three p-orbitals, and so Period 3 of the periodic table holds another eight elements. By the end of the third period, we are up to element 18, argon, because the first three shells can contain 2, 8 and 8 electrons, respectively – a total of 18. In shell 4 (Period 4) a new type of orbital, the **d-orbital**, makes its first appearance, and by the sixth shell, electrons also have an **f-orbital** into which they can fall.

In those shells where they exist, there are three p-orbitals, five d-orbitals and seven f-orbitals. Since each orbital can contain two electrons, there are a possible total of six p-electrons, 10 d-electrons and 14 f-electrons in each of the shells where they occur. Each set of orbitals is also known as a

subshell; if you were building up an atom by adding electrons, as described above, and you had reached shell 4, the order of filling is: s-subshell first, then the d-subshell, then the p-subshell. Similarly, in the sixth shell, the order is s, d, f, p. The structure of the periodic table reflects this order (see page 39); the s-block (Groups 1 and 2) corresponds to the s-subshell (just one orbital); the central block, called the d-block (Groups 3 to 12), corresponds to the d-subshell; the f-block, corresponding to the f-subshell, normally stands apart from the rest of the table, although it is included in the extended version of the table (see page 85), after the d-block; and the right-hand section of the periodic table is the p-block (Groups 13 to 18), which corresponds to the p-subshell.

Unstable nuclei

As we have been filling the electron shells, we should also have been adding protons to the nucleus, since the number of electrons in an atom is equal to the number of protons in the nucleus so that the atom has no overall electric charge. So by now, the nuclei are much bigger than those of hydrogen or helium. Argon, with its 18 electrons, must also have 18 protons in the nucleus. If a nucleus that big consisted only of protons, the protons' mutual repulsion would overpower the attraction of the strong nuclear force. The nucleus would be extremely unstable and would fly apart in an instant. Neutrons provide the attractive strong nuclear force without adding the repulsive electrostatic force: they act like nuclear glue.

So, for example, the most common isotope of argon has 22 neutrons to help its 18 protons

Above: The three 2p orbitals, and an atom with s- and p-orbitals superimposed. In atoms with a filled outer shell, such as neon, the orbitals combine, forming a spherically-symmetrical orbital – such atoms are spheres.

adhere. However, it is not always the case that more neutrons equals greater stability. Certain proton-neutron mixtures are more stable than others, and so for any element some isotopes are more common. The most common isotope of argon is argon-40, with an atomic mass of 40 (the mass of the electron is negligible, so the **atomic mass** is simply the total number of protons and neutrons). However, while argon-40 is by far the most common, there are other stable isotopes. The average atomic mass (the **standard atomic weight**) of any sample of argon atoms is not a whole number: it is 39.948. In fact, no element has a standard atomic weight that is a whole number; chlorine's, for example, is 35.453.

There are several things that can happen to an unstable nucleus. The two most common are alpha decay and beta decay. In **alpha decay**, a large and unstable nucleus expels a clump of two protons and two neutrons, called an alpha particle. The atomic number reduces by two, because the nucleus loses two protons. So, for example, a nucleus of radium-226 (88p, 138n) ejects an alpha particle to become a nucleus of radon-222 (86p, 136n). Alpha decay results in a **transmutation** of one element into another – in this case, radium becomes radon.

This kind of nuclear instability is the reason why there are no more than about 90 naturally-occurring elements. Any heavier ones that were made, in supernovas, have long since disintegrated to form lighter elements. Elements heavier than uranium, element 92, have only been made artificially, and most have only a fleeting existence. For more information on these **transuranium elements**, see pages 153–7. There are two elements with atomic number less than uranium that also have no stable isotopes and are not found naturally: technetium and promethium.

In **beta decay**, a neutron spontaneously changes into a proton and an electron. The electron is expelled from the nucleus at high speed, as a beta particle. This time, the atomic number *increases* by one, since there is now an extra proton in the nucleus. So, while argon-40 is stable, argon-41

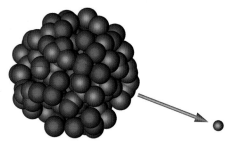

Unstable nucleus increases   fast electron
its atomic number by one   (beta particle)

Above: Beta decay. A neutron in an unstable nucleus spontaneously turns into a proton and an electron. The atomic number increases by one because of the new proton.

(18p, 23n) is not; its nucleus undergoes beta decay to become a nucleus of potassium-41 (19p, 22n). Note how the mass of the nucleus is unchanged – because the new proton has the same mass as the old neutron – despite the fact that the element has transmuted.

Alpha and beta decay are random processes, but in a sample of millions or billions the time for half of them to decay is always the same; this is called the **half-life**.

Nuclear reactions such as alpha and beta decay involve the nucleus losing energy. As a result, the nucleus emits a photon – just as electrons do when they drop down to a lower energy level. But the amount of energy involved in nuclear reactions is much greater, so they produce very energetic **gamma ray** photons rather than photons of visible light or ultraviolet radiation. The disintegration of nuclei, together with the alpha and beta particles and the gamma rays, constitute **radioactivity.**

Bonds
Atoms – with their protons and neutrons in the nucleus and their electrons in orbitals around it – do not exist in isolation. This book alone is composed of countless trillions of them. If you start gathering many atoms of the same element together, the atoms

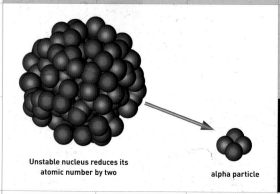

Unstable nucleus reduces its
atomic number by two

alpha particle

Left: Alpha decay. An unstable nucleus loses an alpha particle (2p, 2n), reducing its atomic number by two.

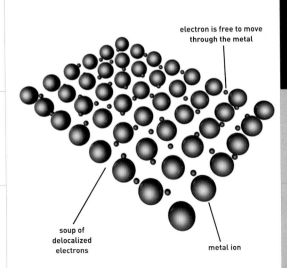

electron is free to move through the metal

soup of delocalized electrons

metal ion

may start to link together, or **bond**, leading to some interesting bulk properties.

Most elements are metals. When metal atoms come together, their outermost electrons become free, or **delocalized**, from their host atoms, so that they are shared between all the metal nuclei. Instead of being allowed only specific energies, they now have a continuous range of possible energies, called a **conduction band**. Because they are free to move, those same electrons are able to absorb almost any photons (of light or other electromagnetic radiation) heading their way, sending the photon back in the same direction from which it came. This is why metals are both opaque and reflective. And because the electrons can move freely, metals are also good **conductors** of electricity. The metallic elements are found on the left-hand side and in the middle of the periodic table.

The electrons in the non-metallic element sulfur, on the other hand, are not free. Instead, they are held in shared orbitals that form the bonds between the atoms. As a result, sulfur is a good **insulator.** Some elements are insulators under normal circumstances, but their electrons can be promoted into a delocalized state and into a conduction band by heat or photons of electromagnetic radiation. Silicon is the best of these **semiconductors**. Non-metals and semi-metals are found to the right of centre in the periodic table.

Some elements do not easily form bonds with atoms of their own kind – in particular, the elements at the extreme right-hand end of the periodic table, which all have completely filled electron shells. These are all gases at room temperature – individual atoms flying around at high speed. They can be forced together to form a liquid or a solid only by cooling them to extremely low temperatures or by exerting extremely high pressures. All other elements that are gases at room temperature exist as small **molecules** of two or three atoms each: for example, hydrogen (H_2), bromine (Br_2), chlorine (Cl_2) and oxygen (O_2 or O_3). The electrons in these molecules are held in **molecular orbitals** that surround all the nuclei involved.

In some cases, a pure sample of an element may take on one of several different forms, depending upon temperature and pressure. Diamond and graphite are both pure carbon, for example. These different forms of the same pure element, with very different properties, are called **allotropes**.

Chemical reactions and compounds

Things get really interesting when atoms of one element interact with atoms of another element. In some cases, the result is a simple **mixture**. Any mixture involving at least one metallic element is called an **alloy**. But in most cases, bonds do actually form between dissimilar atoms, in which case a **chemical reaction** occurs and the result is a **compound**. The chemical reaction involves electrons being either transferred or shared between atoms, to form ionic or covalent bonds, respectively. The result is always a filled outer electron shell, the most stable configuration.

An **ionic bond** involves ions, which form when atoms lose or gain electrons. So, for example, an atom of sodium has just one electron in its outer shell, which it easily loses, thereby attaining a full outer shell. When it does so, the atom has more protons than electrons; the neutral sodium atom has become a positively charged sodium **ion**. Similarly, an atom of chlorine, with seven electrons in its outer shell, easily gains an electron and

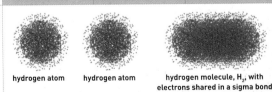

hydrogen atom   hydrogen atom   hydrogen molecule, H_2, with electrons shared in a sigma bond

Formation of sodium chloride, NaCl

Chlorine atom gains an electron to become a chloride ion

Electric force holds sodium and chlorine ions together in a crystal

Sodium atom loses electron easily

The two ions both with filled outer electron shells, have opposite charge

also attains a full outer shell. The neutral chlorine atom becomes a negatively-charged chloride ion. The resulting ions stick together by electrostatic attraction, because they have opposite charges. They form a repeating structure – a crystal of sodium chloride (table salt). Ionic compounds, formed from metal and a non-metal, have high melting points, because ionic bonds are very strong.

Sodium atom $1s^2\ 2s^2\ 2p^6\ 3s^1$
Sodium ion $1s^2\ 2s^2\ 2p^6$
Chlorine atom $1s^2\ 2s^2\ 2p^6\ 3s^2\ 3p^5$
Chloride ion $1s^2\ 2s^2\ 2p^6\ 3s^2\ 3p^6$

A **covalent bond** involves electrons being shared between two or more non-metallic atoms. A molecular orbital forms around the atoms' nuclei. So in the compound methane, for example, made of one carbon atom and four hydrogen atoms (CH_4), covalent bonds form between the carbon and each of the hydrogens.

Compounds with covalently bonded molecules tend to have lower melting points, because the individual molecules are held together more loosely than the atoms in an ionic compound. Water (H_2O), ammonia (NH_3) and carbon dioxide (CO_2) are all covalent compounds. Some covalent compounds have very large molecules. Proteins, for example, typically consist of hundreds or even thousands of atoms.

Some compounds involve a mixture of ionic and covalent bonding. A clump of covalently bonded

atoms can become ionized and can then bond ionically with another ion. This is what happens in calcium carbonate (chalk, $CaCO_3$), in which positively charged calcium ions (Ca^{2+}) bond with carbonate ions ($CO_3{}^{2-}$), which are clumps of covalently bonded carbon and oxygen atoms.

The negatively charged part of an ionic compound is called the **anion** (pronounced "an-ion"), while the positively charged part is called the **cation** (pronounced "cat-ion"). It is the convention in naming ionic compounds that the name of the cation is first and the anion second – so in sodium chloride, the sodium ion is the cation and the chloride ion is the anion.

We have only really scratched the surface of the various ways in which the 90 or so elements combine to make the things around you. But it should be enough to give you a good idea how just three types of particle, plus a bit of quantum weirdness, can give rise to the enormous diversity of substances in the world. It is these substances that are the focus of the rest of this book.

Every element that exists naturally or has been created in laboratories – up to element number 118 – is featured in this book. More space is given to those elements that are particularly important

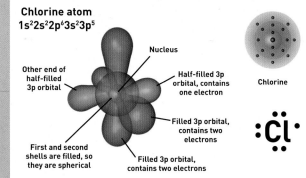

Chlorine atom
$1s^2 2s^2 2p^6 3s^2 3p^5$

Nucleus

Other end of half-filled 3p orbital

Half-filled 3p orbital, contains one electron

Chlorine

Filled 3p orbital, contains two electrons

First and second shells are filled, so they are spherical

Filled 3p orbital, contains two electrons

:C̈l·

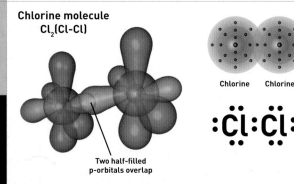

Chlorine molecule
Cl_2 (Cl–Cl)

Chlorine   Chlorine

:C̈l:C̈l:

Two half-filled p-orbitals overlap

Top: Ionic bond. A sodium atom loses its lone outer electron, forming a positive ion. A chlorine atom receives the electron, becoming a (spherically symmetrical) negative ion. Electrostatic attraction then binds the ions together in a cubic crystal (inset).

Right: Example of a covalent bond. Two chlorine atoms, both one electron short of a filled outer shell, contribute one electron each to a bonding orbital formed by two overlapping p-orbitals.

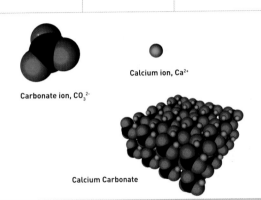

Carbonate ion, CO$_3^{2-}$

Calcium ion, Ca^{2+}

Calcium Carbonate

ATOMIC NUMBER: 17	
ATOMIC RADIUS: 100 pm	
OXIDATION STATES: -1, +1, +2, +3, +4, **+5**, +6, **+7**	
ATOMIC WEIGHT: 35.45	
MELTING POINT: -102°C (-151°F)	
BOILING POINT: -34°C (-29°F)	
DENSITY: 3.20 g/L	
ELECTRON CONFIGURATION: [Ne] 3s^2 3p^5	

Above: Crystal structure of calcium carbonate. Each carbon atom (black) bonds covalently with three oxygen atoms (red), forming a negatively charged carbonate ion. These bond ionically with positively-charged calcium ions.

or interesting. The book is mostly divided according to the vertical columns of the periodic table, called **groups**. Elements that are in the same group have very similar properties, because their outermost electron shell has the same electron configuration. So, for example, lithium (1s^2 2s^1) and sodium (1s^2 2s^2 2p^6 3s^1) both have a single electron in their outermost shell. There are various systems for naming the groups of the periodic table; in this book, we are adopting the one used by the International Union of Pure and Applied Chemistry (IUPAC), in which the groups are numbered from 1 to 18. The elements in the first 12 groups are all metals (apart from hydrogen); the other groups consist of non-metals and metalloids – elements with properties between metals and non-metals.

Some practical issues – the data files
Accompanying the profile of each element in this book is a data file giving that element's **atomic number** and listing several basic physical properties. A sample data file (for chlorine) is included here, to illustrate the properties you are likely to encounter with each element

Atomic weight, (also called relative atomic mass or standard atomic weight) is the mass of a single atom of the element relative to one-twelfth of a carbon-12 atom. It is never a whole number, because elements exist as a mixture of isotopes of different weights.

Atomic radius, in picometres (trillionths of a metre), is not a precise measurement, since electrons exist in fuzzy orbitals. Oxidation state represents the electric charge an atom gains in

forming an ionic compound; in sodium chloride the oxidation state of sodium ions is +1 and chloride ions, -1, since sodium has gained a positive charge and the chloride ions have gained a negative charge. In covalent compounds, the oxidation state is based on how many electrons an atom shares. Many elements can exist in more than one oxidation state. To avoid ambiguity in naming compounds, the oxidation state of a metal ion is often shown in brackets, as a Roman numeral, and is called the oxidation number. So, CuO is copper(II) oxide and Cu$_2$O is copper(I) oxide.

Melting point and boiling point are given in degrees Celsius (the unit many people still call "centigrade") and in Fahrenheit, and are as measured at average atmospheric pressure. Scientists normally use degrees Kelvin; the Kelvin scale starts at the coldest possible temperature (absolute zero), which is –273.15°C (–459.67°F). Since it is not familiar in everyday use, the Kelvin scale is not used in the data tables of this book.

The density of an element is simply the mass of a sample of the element divided by its volume. In this book, density is given in units of grams per cubic centimetre (g/cm^3) for solids and liquids, and in grams per litre (g/L) for gases. Density depends upon temperature; solid and liquid densities are their values at room temperature, gases at 0°C.

Electron configuration is the arrangement of the electrons, in shells and subshells (see above). Only the outer electron shell is shown; the filled inner shells are represented by the relevant element from Group 18 of the periodic table. Chlorine's electron configuration in full is 1s^2 2s^2 2p^6 3s^2 3p^5; but the first two filled shells are the same as those for the noble gas neon (Ne). So, the shortened version of chlorine's electron configuration reads: [Ne] 3s^2 3p^5.

Elements – a history

Ancient peoples were familiar with several of the substances that we now know as chemical elements. Some, such as gold, silver and sulfur, exist naturally in a relatively pure form; others, such as iron, copper and mercury, are easily extracted from minerals. But it was not until the end of the eighteenth century that scientists established the notion of what a chemical element actually is, and how that differs from a chemical compound. And it was the 1920s before all the naturally occurring elements had been discovered and isolated.

Trying to make sense of the incredible diversity of matter must have been bewildering for ancient philosophers. In many early civilizations, philosophers deduced that all matter is made up of earth, air, fire and water, in varying mixtures. These were the "elements" as the ancients understood them [note: that would have made this book a bit shorter!]. Transformations of matter – what we now call chemical reactions – were believed to be changes in the amounts of those elements present in a substance.

Notions of the four classical elements formed the basis for the mystical art of alchemy, whose best-known aim was the transformation of "base metals" such as lead into gold. Alchemy was as practical as it was mystical, and many of the basic techniques used by chemists to this day were developed by alchemists. And although the theories of alchemy turned out to be false, the alchemists in Ancient China, in the Islamic Caliphate and in medieval Europe built up a working knowledge of many important chemical substances and their reactions. It was not only the alchemists who helped gather practical knowledge about matter and chemical reactions: for example, early apothecaries (pharmacists), glassmakers and, perhaps most importantly, metallurgists also contributed know-how and experience. In early modern Europe, a new version of alchemy developed, with mercury, sulfur and salt at its core – but the focus was on the "principles" of these substances, rather than their physical properties.

Inevitably, the flaws in the theories of alchemy were exposed by the scientific method, which became popular in Europe in the seventeenth century. Crucially, chemists showed that air is a mixture of gases, so it cannot be an element, and that water is a compound.

Anglo-Irish scientist Robert Boyle's book *The Sceptical Chymist*, published in 1661, encouraged scientists to question the accepted alchemical explanations and to take a rigorous scientific approach to working out what the world is made of. Boyle promoted the use of chemical analysis, a systematic approach by which chemists could determine the component substances in a mixture

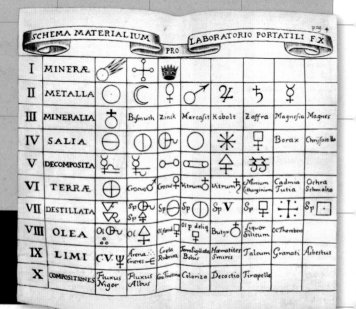

Right: Table from *Opuscula Chymica* (1682), by the German polymath Joachim Becher. The table is an attempt at classifying the known substances under various categories. Like Boyle in England, Becher was an alchemist who thought scientifically.

Left: French chemist Antoine Lavoisier, often referred to as "the father of modern chemistry".

Below: Lavoisier's list of chemical elements – his new names for them on the left and their old names on the right. The first two "elements" are light and heat (lumière and calorique). The list also includes chalk (chaux), now known to be a compound.

	Noms nouveaux.	Noms anciens correspondans.
	Lumière.........	Lumière.
Substances simples qui appartiennent aux trois règnes & qu'on peut regarder comme les élémens des corps.	Calorique.......	Chaleur. / Principe de la chaleur. / Fluide igné. / Feu. / Matière du feu & de la chaleur.
	Oxygène........	Air déphlogistiqué. / Air empiréal. / Air vital. / Base de l'air vital.
	Azote...........	Gaz phlogistiqué. / Mofete. / Base de la mofete.
	Hydrogène.......	Gaz inflammable. / Base du gaz inflammable.
Substances simples non métalliques oxidables & acidifiables.	Soufre..........	Soufre.
	Phosphore.......	Phosphore.
	Carbone.........	Charbon pur.
	Radical muriatique.	Inconnu.
	Radical fluorique .	Inconnu.
	Radical boracique,.	Inconnu.
Substances simples métalliques oxidables & acidifiables.	Antimoine.......	Antimoine.
	Argent..........	Argent.
	Arsenic.........	Arsenic.
	Bismuth.........	Bismuth.
	Cobolt..........	Cobolt.
	Cuivre..........	Cuivre.
	Etain...........	Etain.
	Fer	Fer.
	Manganèse.......	Manganèse.
	Mercure.........	Mercure.
	Molybdène.......	Molybdène.
	Nickel..........	Nickel.
	Or..............	Or.
	Platine.........	Platine.
	Plomb..........	Plomb.
	Tungstène.......	Tungstene.
	Zinc...........	Zinc.
Substances simples salifiables terreuses.	Chaux..........	Terre calcaire, chaux.
	Magnésie........	Magnésie, base du sel d'Epsom.
	Baryte..........	Barote, terre pesante.
	Alumine........	Argile , terre de l'alun, base de l'alun.
	Silice..........	Terre siliceuse, terre vitrifiable.

or compound. A new breed of chemists heeded Boyle's advice, and in the eighteenth century – thanks to alternative theories, rigorous testing and open minds – the new science of chemistry began to take great strides forwards.

In his influential book, Robert Boyle expounded an idea that was gaining popularity at the time and that was crucial to the development of modern chemistry: that matter is made of countless tiny particles. Many philosophers had considered the idea, even in ancient times, but Boyle was the first person to connect particles with elements, compounds and chemical reactions. He even suggested that elements are made of particles that are "primitive and simple, or perfectly unmingled" that are "the ingredients" of compounds.

The concept of an element came into sharp focus with the insight of French chemist Antoine Lavoisier. In his 1789 book *Traité élémentaire de chimie* ("Elementary Treatise on Chemistry"), Lavoisier proposed that an element should be defined simply as a substance that cannot be decomposed.

Lavoisier's insight into chemical elements was largely a result of his careful quantitative experiments: he carefully weighed the reactants and products in a range of chemical processes, and proved that no mass is lost during chemical reactions. Crucially, he studied reactions in closed vessels so that gases absorbed or released during reactions were included in his calculations. When one substance reacts with another, they simply combine to make a third one – and that product of the reaction can be decomposed into its simpler components. Lavoisier's master stroke was in explaining combustion (burning) as the combination of substances with oxygen. He worked out that when hydrogen burns in air, it combines with oxygen to make water; he even managed to decompose water into its two constituent elements.

In 1808, English chemist John Dalton united Lavoisier's understanding of elements and compounds with Boyle's insistence on the particulate nature of matter. In his book *A New System of Chemical Philosophy*, Dalton proposed that all the atoms of a particular element are identical and different from those of other elements. The crucial, and measurable, difference was the atoms' masses: hydrogen atoms are the lightest, oxygen heavier, sulfur heavier still and iron even heavier. This made sense of the fact that compounds are always composed of fixed ratios of substances by mass. For example, the iron in a sample of iron sulfide always accounts for 63 per cent of the compound's mass, however large or small the sample.

The rise of the scientific approach to chemistry led to the discovery of several new elements in the eighteenth century – and Lavoisier's definition of elements, and his insight into the role of oxygen in combustion, helped to speed up the rate of discovery in the nineteenth century. Many new metals were isolated from their "earths" (oxides) by removing oxygen. The invention of the electric battery in 1799 gave chemists a new tool for chemical analysis. Electric current can decompose a compound that resists most other forms of analysis. Several new elements were discovered using electrolysis ("electrical splitting") in the first three decades of the nineteenth century. In the 1860s, two German scientists, Robert Bunsen and Gustav Kirchhoff, added another important technique to the practice

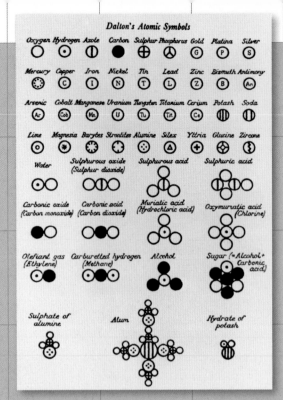

of analytical chemistry: spectroscopy. Using an instrument they invented – the spectroscope – Bunsen and Kirchhoff studied the spectrum of light given out when particular elements are vaporized and heated. They studied the emission spectra of all the known elements – and the presence of unfamiliar lines in the spectra of various substances led to the discovery of several previously unknown elements.

By this time, chemists began to realize that the growing list of elements seemed to fall into groups according to their properties and reactions. Sodium, potassium and lithium are all metals that react with water to produce alkaline solutions; chlorine, bromine and iodine all react with metals to make compounds like common salt. English chemist John Newlands noticed that elements in the same group seemed to be spaced eight elements apart in a list of the elements by atomic weight. Newlands' scheme only worked for the first 20 or so elements, and other chemists

Top: Illustrations from Dalton's book *A New System of Chemical Philosophy* (1808), suggesting how elements and compounds might relate to atoms and molecules.

Left: English chemist John Dalton.

ridiculed him. However, Russian chemist Dmitri Mendeleev found similar "periodicity" when he organized the known elements into groups, based on their properties and reactions. In 1869, Mendeleev formulated the first periodic table, revealing at last a sense of order in the growing list of elements.

One of Mendeleev's master strokes was leaving gaps in his table, where elements would fit when and if they were discovered. From their positions in the periodic table, Mendeleev could predict the missing elements' atomic weights and chemical properties. Within a few years, several of the missing elements had been found.

The discoveries of the electron, radioactivity and X-rays in the 1890s were the stimulus for a dramatic era of atomic physics in the first half of the twentieth century. The existence of small, light negatively charged electrons showed for the first time that atoms had inner structure. The behaviour of electrons explained how atoms can form ions, how atoms form bonds and why some elements are more reactive than others. Using radioactivity, in 1911 New Zealand-born physicist Ernest Rutherford discovered the atomic nucleus – a tiny, dense, positively charged particle at the centre of each atom – and proposed that negatively charged electrons orbit the nucleus. Dutch physicist Antonius van den Broek originated

Above: Dmitri Mendeleev.

Bottom: Dmitri Mendeleev's first periodic table (1869), with elements listed by atomic weight and sorted into groups, which unlike the modern table are horizontal. Question marks represent then-unknown elements whose existence Mendeleev predicted.

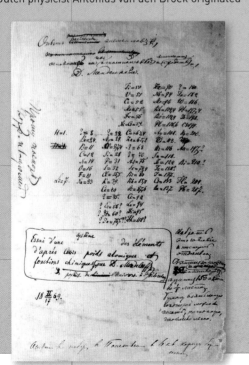

the concept of atomic number when he worked out that each element had a different number of positive charges in the nucleus, corresponding to the number of electrons in orbit around it. In 1917, Rutherford discovered that the nucleus was made of particles, which he named protons.

In 1913, Danish physicist Niels Bohr used the nascent theory of quantum physics to work out that electrons travel around the nucleus only in certain specific orbits, and found that changes in energy of electrons moving between the various orbits matched the spectra Bunsen and Kirchhoff had studied. The very highest energy transitions emit X-rays, not visible light or ultraviolet; in 1914, English physicist Henry Moseley found a correspondence between the positive charge of an element's nucleus and its X-ray spectrum. This allowed him to refine the periodic table – arranging elements precisely by atomic number, rather than atomic weight – and to predict two more unknown elements. The discovery of the neutron in 1932 completed the basic understanding of atoms, including an explanation of the occurrence of isotopes – different versions of the same element, with differing numbers of neutrons in the nucleus (see page 8). The theories and experiments of nuclear physics enabled physicists to work out how elements are created, by the fusion of protons and neutrons inside stars and supernovas. Nuclear physics also led to the creation of elements heavier than uranium, most of which do not exist naturally (see pages 153–7).

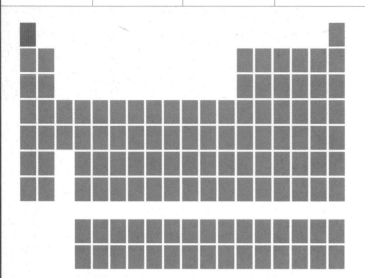

1
H
Hydrogen

Hydrogen

Hydrogen is the most abundant of all the elements, constituting more than 75 per cent of all ordinary matter in the Universe by mass (most of the mass of the Universe is "dark matter", whose nature remains a mystery), and accounting for around 90 per cent of all atoms. Most of the hydrogen on Earth is in water molecules, but this element is also a crucial component in the molecules involved in the processes of life. Hydrogen may even replace fossil fuels as the main energy source in the future.

Hydrogen is officially in Group 1 of the periodic table, but it is so different from the other Group 1 elements that it is generally considered in a category of its own. The single electron of a hydrogen atom half-fills an s-orbital around the nucleus (see page 8), as do the outermost electrons in the other elements of Group 1. And, like those other elements, a hydrogen atom will readily lose its electron, becoming a positive ion, H+. However, a hydrogen atom will also readily accept an electron, so that it has a full shell. In that case it becomes a negative hydrogen ion, H–, in the same way as the elements of Group 17.

In another break with the properties of the rest of Group 1, hydrogen is a gas (H_2) at room temperature; all the other Group 1 elements are solid metals. However, in the extreme pressures at the centre of gas giant planets such as Jupiter, hydrogen does behave like a metal. The vast clouds of dust and gas from which stars are born are mostly hydrogen. Wherever it is irradiated by radiation from nearby stars, it produces a beautiful reddish-pink glow. This is due to the electrons in countless hydrogen atoms being kicked up to a higher energy level and then falling back down, emitting photons as they do so. The reddish light is due to a common transition, from energy level n=3 down to n=2 (see page 7). Astronomers observe this "hydrogen alpha" radiation coming from gas clouds in every corner of the Universe.

Left: Two enormous nebulas – clouds of gas and dust – in the constellation Cygnus. The red light they emit is hydrogen-alpha, produced as electrons jump between energy levels in hydrogen atoms.

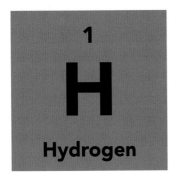

1

H

Hydrogen

ATOMIC NUMBER: 1

ATOMIC RADIUS: 30 pm

OXIDATION STATES: -1, +1

AVERAGE ATOMIC WEIGHT: 1.008 g mol^{-1}

MELTING POINT: -259.14°C (-434.45°F)

BOILING POINT: -252.87°C (-423.17°F)

DENSITY AT STP: 0.09 g/L

ELECTRON CONFIGURATION: 1s^1

English scientist Henry Cavendish is normally given credit for the discovery of hydrogen, after he produced and studied it in 1766. The gas Cavendish produced was explosive, and Cavendish suggested that it might be rich in a hypothetical substance that scientists of the day named "phlogiston". However, Cavendish couldn't explain why his "phlogisticated air" produced water when it burned. French chemist Antoine Lavoisier found the explanation in 1792, and named the element *hydrogene*, from the Greek for "water generator".

The element hydrogen exists as a gas at normal temperature and pressure. Hydrogen gas (H_2), also called dihydrogen, is composed of molecules, each made up of two hydrogen atoms. It is found in ordinary air, albeit in tiny quantities, making up less than one-millionth of the atmosphere. This is mainly because hydrogen molecules are so light that they escape into space.

On Earth, most hydrogen is combined with oxygen, in water molecules, H_2O. As a result, more than 10 per cent of the mass of any ocean is hydrogen, despite this element's very low atomic mass. Water is a great solvent, dissolving most substances at least to some extent. The reason for this is that water molecules can easily separate, or dissociate, into H^+ and OH^- ions, and these ions can hold on to other ions by electrostatic attraction.

Acidic solutions have greater concentrations of hydrogen ions (H^+) than does pure water. The measure of the acidity of a solution, known as pH, is actually a measure of the concentration of H^+ ions in that solution. Acids react energetically with most metals: the metal atoms dissolve in the acid, displacing the hydrogen ions and forcing

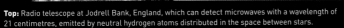

Top: Radio telescope at Jodrell Bank, England, which can detect microwaves with a wavelength of 21 centimetres, emitted by neutral hydrogen atoms distributed in the space between stars.

Above: Bubbles of hydrogen gas being produced by a reaction between zinc metal and hydrochloric acid.

Right: Computer visualization of a water molecule. The colours represent the distribution of the molecule's bonding electrons. The electron density is less around the hydrogen atoms, so those parts of the molecule have a slight positive charge (red).

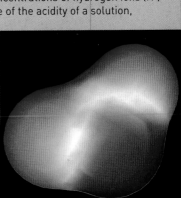

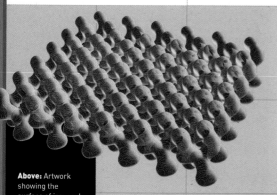

Above: Artwork showing the surface of ice; each blue particle is a water molecule. The attraction between water molecules is strengthened by hydrogen bonds between adjacent water molecules.

Above right: The destruction of the hydrogen-filled airship *LZ 129 Hindenburg*, at Lakehurst Naval Air Station, New Jersey, USA, on 6 May 1937.

Below: Margarine containing hydrogenated vegetable oils. As a result of health concerns, most margarines are now made with vegetable oils blended with buttermilk, rather than with hydrogenated vegetable oils.

them to pair up to produce molecules of hydrogen gas. Several scientists had produced hydrogen in this way before it was realized that hydrogen is an element.

In addition to the hydrogen found in the water molecules they contain, living things also have hydrogen in every organic molecule, including proteins, carbohydrates and fats. The presence of hydrogen atoms is crucial in large organic molecules, giving structure and stability through a special type of bond called the hydrogen bond. The double helix of DNA (deoxyribonucleic acid) relies upon hydrogen bonds, which are strong enough to keep the two strands of the double helix together, but weak enough that the strands can be separated during replication of the DNA for cell division in growth and reproduction.

Hydrogen bonding is also found in water, and it results in a greater attraction between water molecules than would otherwise be the case. Without hydrogen bonding, water would boil and freeze at much lower temperatures.

Fossil fuels, such as oil, coal and natural gas, consist mostly of hydrocarbons – molecules containing only carbon and hydrogen. When fossil fuels burn, oxygen atoms combine with the hydrocarbons, producing carbon dioxide (CO_2) and water (H_2O). Natural gas is the main source of hydrogen for industry. In a process called steam reforming, superheated steam separates the hydrogen from hydrocarbons such as methane (CH_4).

Almost two-thirds of all industrially produced hydrogen is used to make ammonia (NH_3), around 90 per cent of which in turn is used in the manufacture of fertilizers. Most of the rest of the hydrogen supply is used in processing crude oil, to help "crack" large hydrocarbon molecules into smaller molecules needed in commercial fuels and rid hydrocarbon molecules of unwanted sulfur atoms.

In the first few decades of the twentieth century, hydrogen was produced in large quantities for use in airships. The gas is much less dense than air, and easier and cheaper to produce than helium. However, the high combustibility of hydrogen caused a number of horrific accidents, most notably the tragic explosion that destroyed the German transatlantic airship *LZ-129 Hindenburg* in 1937. Thirty-six people died when a million litres of hydrogen in the airship's huge envelope caught fire on arrival in New Jersey, USA.

Since the early twentieth century, hydrogen has been used in large quantities to produce fats for the food industry, by hydrogenating cheap liquid vegetable oils. The resulting "trans fats" are solids at room temperature, and have a longer shelf life than the liquid oils. However, starting in the 1950s, research has found that trans fats increase the risk of cancers and heart disease; as a result, the use of hydrogenated fats is now heavily regulated and is on the decline.

There are three isotopes of hydrogen. The most common, with a single proton as its nucleus, is referred to as protium. The only other stable isotope is deuterium (D), which has one proton and one neutron. Deuterium is also called heavy hydrogen, and water made with deuterium (D_2O), called heavy water, is more than 10 per cent denser than ordinary water. The third isotope, tritium, has a proton and two neutrons. It decays through beta decay (see page 9), and has a half-life of 12.3 years.

Deuterium and tritium are involved in experiments with nuclear fusion, which could provide a practically limitless supply of energy in the future. In most fusion reactors, deuterium nuclei (1p, 1n) and tritium nuclei (1p, 2n) come together at extremely high temperatures and join (fuse) to produce helium-4 nuclei (2p, 2n); a neutron (n) is released as a result of each helium nucleus created. The reaction unleashes enormous amounts of energy. In all experiments so far conducted, the amount of energy used to start the reaction exceeds the amount produced. But nuclear technologists hope that within 20 or 30 years, fusion reactors might become economically viable and reduce our reliance on fossil fuels and conventional (fission) nuclear power. Fusion reactions involving deuterium and tritium are also the source of energy of hydrogen bombs. Inside an H-bomb, a conventional atomic bomb creates sufficiently high pressure and temperature for fusion to occur.

Even before nuclear fusion becomes viable, hydrogen may replace fossil fuels as a common energy "currency". The need to cut carbon dioxide emissions, together with the fact that fossil fuel reserves are limited, means that our reliance on fossil fuels cannot last forever. Burning hydrogen produces only water as a waste product, and hydrogen is plentiful and easy to produce. Of course, energy is needed in the first place to produce the hydrogen; electricity from renewable sources can be used to separate it from water, through a process called electrolysis. The resulting hydrogen has a high energy density and can be stored and transported fairly easily. Most hydrogen-powered vehicles are powered by hydrogen fuel cells, which rely upon a chemical reaction that is the reverse of electrolysis: hydrogen combines with oxygen. The reaction is the equivalent of burning hydrogen – the waste product is water – but is slower and more controlled, and produces electrical energy instead of heat.

Above: A hydrogen-powered vehicle being refuelled with hydrogen at an experimental filling station. Inside the car, a fuel cell produces electrical power from the reaction of hydrogen with oxygen.

Below: Explosion of the George device, part of a series of nuclear tests conducted by the USA in 1951, in the Marshall Islands in the Pacific Ocean. George was the first bomb in which nuclear fusion was achieved.

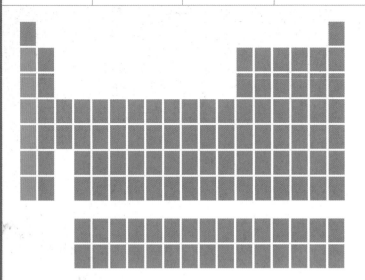

		3
		Li
		Lithium

		11
		Na
		Sodium

		19
		K
		Potassium

		37
		Rb
		Rubidium

		55
		Cs
		Caesium

		87 ☢
		Fr
		Francium

The Alkali Metals

In contrast with the uniqueness of hydrogen and the varied properties you find in some other groups of the periodic table, the alkali metals exhibit particularly strong family resemblances. All are solid, but soft, at room temperature; all are shiny metals that have to be kept under oil or in inert atmospheres because they are very reactive.

With the exception of the radioactive and extremely rare francium, these shiny metals react rapidly with oxygen in the air, so that their surfaces quickly become dull. Their shininess is revealed again if you cut them with a knife. Caesium is the most chemically reactive of all the elements in this group, and spontaneously catches fire in air.

The atoms of these elements all have just one electron in their outer shell (s^1) – and this is why they are so reactive. Atoms of alkali metals only have to lose a single electron to achieve a stable filled shell configuration (see page 11). As they do so, the atoms transform into positive ions. As a result, alkali metals readily form ionic compounds – in particular with the elements of Group 17, whose atoms are only one electron short of being filled so make the perfect partners for alkali metals. One of the most familiar examples of this is sodium chloride (common salt).

All Group 1 elements react violently in water (H_2O), displacing hydrogen ions (H^+) to produce hydrogen gas, leaving an excess of hydroxide ions (OH^-) in solution in the remaining water. Any solution with more OH^- than H^+ ions is alkaline, hence the name of the group: dissolving these elements in water results in a strongly alkaline solution.

3
Li
Lithium

ATOMIC NUMBER: 3
ATOMIC RADIUS: 145 pm
OXIDATION STATES: +1
ATOMIC WEIGHT: 6.94
MELTING POINT: 180°C (357°F)
BOILING POINT: 1,345°C (2,448°F)
DENSITY: 0.53 g/cm³
ELECTRON CONFIGURATION: [He] 2s¹

Lithium is the least dense solid element, and one of the most reactive metals. It was discovered in 1817, by 25-year-old Swedish chemist Johan Arfwedson, who was actually searching for compounds of the recently discovered element potassium in a sample of a translucent mineral called petalite. Arfwedson based the name of the new element on the Greek word *lithos*, meaning "stone". The mineral arfedsonite is named after him – although, ironically, it contains no lithium. English chemist Humphry Davy was the first to extract atoms of the pure element, by electrolysis, in 1818.

The pure element is extracted by electrolysis of the compound lithium chloride (LiCl). Like most other lithium compounds used today, lithium chloride is produced from lithium carbonate (Li₂CO₃), which is produced from rocks containing lithium. Compounds of lithium have many uses, most importantly in lithium ion rechargeable batteries, commonly found in cameras and laptop computers. These batteries are also used in electric vehicles; demand for the element is set to rise dramatically as electric cars become more commonplace.

Lithium compounds are used in the glass and ceramics industry, to reduce the melting point of the ingredients of glass and to increase heat resistance in glass and ceramic cookware. Lithium carbonate is the active ingredient in medicines for various psychological disorders, and is also an important ingredient in the extraction of aluminium. Lithium hydroxide, prepared by driving off carbon dioxide from lithium carbonate, is used as a "scrubber", to absorb carbon dioxide from the air in spacecraft and submarines.

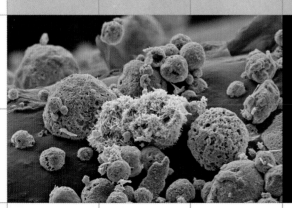

Above: Elemental lithium – a shiny metal that is soft enough to be cut with a knife. The surface of the metal reacts slowly with oxygen from the air to form a dull grey layer of lithium oxide and hydroxide.

Left: False colour scanning electron micrograph of crystals used in lithium-ion rechargeable batteries. During discharging, lithium ions (Li⁺) become inserted (intercalated) into these crystals; during charging, they leave.

Right: Characteristic red light produced by a lithium compound in a flame. The light is produced by electrons that, after being excited by heat, drop to a lower energy level.

11

Na

Sodium

ATOMIC NUMBER: 11

ATOMIC RADIUS: 180 pm

OXIDATION STATES: +1

ATOMIC WEIGHT: 22.99

MELTING POINT: 98°C (208°F)

BOILING POINT: 883 °C (1,621°F)

DENSITY: 0.97 g/cm^3

ELECTRON CONFIGURATION: [Ne] 3s^1

Humphry Davy discovered sodium in 1807, by passing an electric current through molten caustic soda (sodium hydroxide, NaOH). The name of the element is derived from *sodanum*, the Roman name for glasswort, a genus of plants whose ashes were once used in glassmaking. Glassworts are halophytes – salt-loving plants – and their ashes contain sodium carbonate, or soda lime, still an important ingredient in glassmaking. Soda lime glass is used to make bottles and window panes; around 2 kilograms of sodium carbonate is used for every 10 kilograms of glass.
The chemical symbol for sodium, Na, comes from *natrium*, the Latin name for sodium carbonate. In Ancient Egypt, powdered sodium carbonate, called natron, was used as a drying agent in mummification.

Sodium is the sixth most abundant element in Earth's crust, and there are more than 10 kilograms of it, as dissolved sodium ions (Na$^+$), in every cubic metre of seawater. Elemental sodium is produced industrially by electrolysis of molten sodium chloride (NaCl). About 100,000 tonnes of pure sodium are produced each year.

Liquid sodium is used as a coolant in some nuclear power stations. Pure sodium is also used in the manufacture of sodium lamps. The most common type, the low-pressure sodium lamp, is commonly used in street lighting. Inside the lamp's glass bulb is a small amount of solid sodium. After the lamp is turned on, the sodium vaporizes and the lamp emits a characteristic vivid orange glow, produced by electrons dropping from a higher to a lower energy level in the sodium atoms (see page 7).

Most industrial sodium compounds are produced from sodium chloride (NaCl). More than 200 million tonnes of sodium chloride are produced worldwide every year, most of it obtained from rock salt in underground mines. Rock salt is used in its natural state, but crushed, to grit roads in winter. Sodium chloride is normally

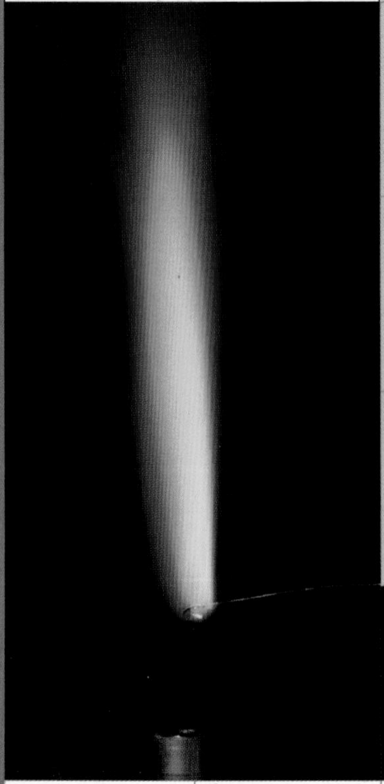

Above: Sodium metal, a soft, silvery metal similar to lithium, which will react violently with water, producing hydrogen gas.

Left: Characteristic orange light produced by a sodium compound in a flame. The light is produced by electrons that, after being excited by heat, drop to a lower energy level.

extracted from rock salt deposits by pumping warm water underground, and then pumping out and evaporating the resulting brine. It has many uses, including food preservation and seasoning (as table salt).

Sodium is essential for proper functioning of nerve cells, and it is one of the most important electrolytes – dissolved ions that help to regulate the body's level of hydration. Too much salt (sodium chloride) can cause the body to retain too much fluid, however, which raises blood pressure – but the long-term effects of high salt intake are unclear. In most countries, the law demands that food manufacturers display the amount of salt added to processed foods, often also giving a "sodium equivalent"; there are 0.4 grams (400 milligrams) of sodium in every gram of salt.

Sodium bicarbonate ($NaHCO_3$) is another sodium compound used in the food industry; it is used as a raising agent because it breaks down with heat or acids, producing carbon dioxide. Sodium hydroxide is a strong alkali with many uses in industry, with important roles in paper-making and the extraction of aluminium, for example; more than 60 million tonnes are produced annually. It is also one of the main ingredients of soap. Sodium carbonate (Na_2CO_3), as washing soda, is also used as a water softener and a descaler.

Above left: Low-pressure sodium street lamps. Electrons produced at the lamp's negative terminal cross the low pressure sodium vapour inside the glass bulb, exciting any sodium atoms with which they collide.

Above right: Grains of table salt (sodium chloride). Each one contains more than a billion billion sodium ions and the same number of chloride ions, ionically bonded in a cubic crystal structure.

Below: Salt pans in Gozo, Malta, that have been used for several thousand years to extract salt from seawater by evaporation.

19

K

Potassium

ATOMIC NUMBER: 19
ATOMIC RADIUS: 220 pm
OXIDATION STATES: +1
ATOMIC WEIGHT: 39.10
MELTING POINT: 63°C (146°F)
BOILING POINT: 760°C (1,398°F)
DENSITY: 0.86 g/cm³
ELECTRON CONFIGURATION: [Ar] 4s¹

Potassium was first isolated by English chemist Humphry Davy; it was the first element that he discovered. In 1807, he passed electric current through molten caustic potash (potassium hydroxide, KOH), and noticed that the silvery-grey particles of potassium "skimmed about excitedly with a hissing sound, and soon burned with a lovely lavender light".

Davy named potassium after the compound he had used in its discovery, potash. That compound, in turn, was named after the ashes left in pots in which people burned plants such as bracken. The ashes were used to make soaps – and, along with sodium hydroxide, potassium hydroxide is still used in soap-making today. Soaps made with potassium tend to dissolve more readily in water, so liquid soaps are normally made with potassium hydroxide, while solid soaps are made with sodium hydroxide. The element's symbol, K, is from the Latin word for alkali: *kalium*.

Potassium is the seventh most abundant element in Earth's crust. The pure metal is produced industrially by heating potassium chloride (KCl) from potassium-bearing minerals with pure sodium vapour. The sodium displaces the potassium, forming sodium chloride and releasing pure potassium as a vapour. Only about 200 tonnes of the metal is extracted each year; as with sodium, industrial use of potassium involves potassium's compounds, rather than the element itself. This is just as well, because potassium is very reactive, and transporting it is difficult and costly.

The three most important potassium compounds are potassium chloride (KCl), potassium nitrate (KNO$_3$) and potassium hydroxide (KOH). More than 90 per cent of the potassium chloride and potassium nitrate produced goes into fertilizers. Most fertilizers are based upon a mix of three elements – nitrogen, phosphorus and

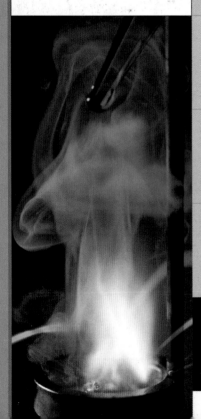

Above: A sample of potassium, a softy, shiny metal. Visible around the edges is a layer of tarnish, composed of potassium oxide and hydroxide, formed by reaction of the potassium with water and oxygen from the air.

Left: Potassium metal reacting violently with water being dripped on to it from above. The reaction produces heat, which causes the potassium atoms to emit lilac-coloured light, characteristic of the element.

potassium, all of which are essential to plant health and growth. Potassium nitrate, from bird droppings (guano) or the mineral saltpetre, was traditionally used in the recipe for gunpowder. As well as its use in soaps, and many uses in the chemical industry, potassium hydroxide is used in alkaline batteries, including rechargeable nickel-cadmium (NiCad) and nickel metal hydride (NiMH) ones.

In humans and other animals, potassium is key to the transmission of nerve impulses, and is an important electrolyte, like sodium. The average adult human body contains about 140 grams of potassium, mostly as dissolved potassium ions (K^+) inside red blood cells.

All fruits and vegetables contain plenty of potassium, and vegans and vegetarians tend to have a higher intake than meat eaters. A medium-sized banana contains about 400 milligrams of potassium: about five thousand million million million potassium atoms.

One in every ten thousand or so potassium atoms is the isotope potassium-40, which is unstable and undergoes beta decay (see page 9), with a half-life of around a billion years. In the average banana, about 15 potassium-40 nuclei decay every second, producing 15 energetic beta rays. Inside the body, thousands of potassium nuclei decay every second, and the resulting beta rays can damage DNA. Fortunately, cells have a built-in DNA repair kit, which can deal with most of this kind of damage.

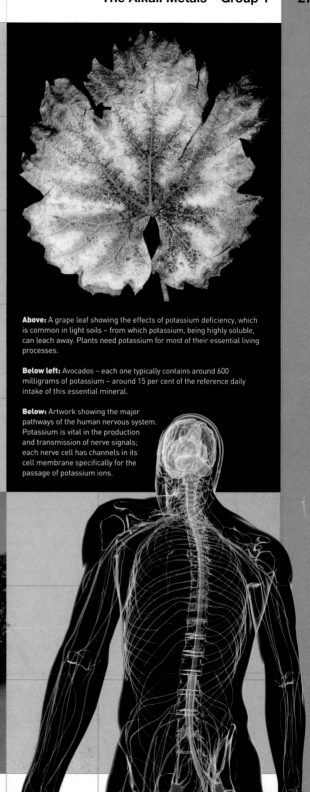

Above: A grape leaf showing the effects of potassium deficiency, which is common in light soils – from which potassium, being highly soluble, can leach away. Plants need potassium for most of their essential living processes.

Below left: Avocados – each one typically contains around 600 milligrams of potassium – around 15 per cent of the reference daily intake of this essential mineral.

Below: Artwork showing the major pathways of the human nervous system. Potassium is vital in the production and transmission of nerve signals; each nerve cell has channels in its cell membrane specifically for the passage of potassium ions.

37

Rb

Rubidium

ATOMIC NUMBER: 37
ATOMIC RADIUS: 235 pm
OXIDATION STATES: +1
ATOMIC WEIGHT: 85.47
MELTING POINT: 39 °C (103°F)
BOILING POINT: 688 °C (1,270°F)
DENSITY: 1.63 g/cm³
ELECTRON CONFIGURATION: [Kr] 5s¹

The German chemists Robert Bunsen and Gustav Kirchhoff discovered the element rubidium in 1861, within a year of their discovery of another Group 1 element, caesium (see below). In both cases, they became aware of the existence of a previously unknown element by the coloured lines in the spectrum produced by samples when they were heated in a flame. The two scientists had been cataloguing the spectra of the known elements, using the laboratory gas burner Bunsen had invented. Bunsen named the element rubidium because of two deep red lines in its spectrum. The pure element was not extracted until 1928.

Only about 3 tonnes of pure rubidium are produced each year, mostly as a by-product of lithium extraction. Rubidium melts at 39.3°C; it can become molten on a hot summer day. Although rubidium is a fairly abundant element in Earth's crust, it has few applications and is not essential in living organisms – although the human body will absorb it because of its similarity to potassium. The radioactive isotope rubidium-87 is used in medicine: it is absorbed into blood cells and is particularly easy to detect by magnetic resonance imaging (MRI), enabling radiographers to pinpoint regions of low blood flow (ischaemia).

Rubidium compounds are used in some solar cells, and in the future this element may be put to use in ion drive engines for spacecraft exploring deep space, since rubidium ionizes very easily.

Above: Rubidium turns the bunsen flame a distinctive red-violet when held in it.

Left: Elemental rubidium, a shiny metal with a melting point of 39°C – lower than that of candle wax and only slightly higher than that of chocolate.

55

Cs

Caesium

Unlike other elements in Group 1, pure caesium has a slight golden tinge. It was the first of the two elements discovered by Bunsen and Kirchhoff, this time from compounds extracted from spring water, in 1861. The name is from the Latin *caesius*, meaning "sky blue". With a melting point of 28.4°C, caesium can become liquid in a warm room.

One of the most important and unusual applications of caesium is its use in atomic clocks. At the heart of these incredibly accurate devices is a cavity in which caesium atoms are excited by microwaves. At a particular microwave frequency, electrons in the caesium atoms produce radiation of exactly the same frequency, and the cavity is said to be resonant. The microwave frequency

ATOMIC NUMBER: 55
ATOMIC RADIUS: 245 pm
OXIDATION STATES: +1
ATOMIC WEIGHT: 132.91

MELTING POINT: 28 °C (83°F)
BOILING POINT: 671 °C (1,240°F)
DENSITY: 1.87 g/cm³
ELECTRON CONFIGURATION: [Xe] 6s¹

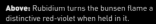

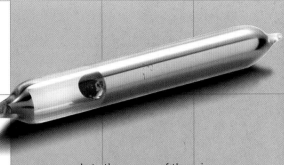

corresponds to the energy of the microwave photons that caesium produces, which is determined by the electron energy levels – ultimately by the laws of nature – so this is a very accurate and reliable way of measuring time. The official world time – Coordinated Universal Time – is the average time kept by more than 70 atomic clocks in laboratories across the world. The most accurate can tell the time so reliably that two of them would differ by no more than a second after 100 million years. Because of its use in atomic clocks, caesium is at the heart of the definition of a second. In 1967, the International Committee on Weights and Measures (CIPM) decided that "The second is the duration of 9 192 631 770 periods of the radiation corresponding to the transition between the two hyperfine levels of the ground state of the caesium-133 atom".

Above left: Elemental caesium, a shiny metal with a melting point of 28°C – the lowest of any metallic element apart from mercury (and francium, which is exceedingly scarce).

Above right: The vacuum chamber of a caesium atomic clock, at the National Physical Laboratory (NPL) in the UK. The world's official time – Coordinated Universal Time – is based on the time kept by a number of caesium atomic clocks, including one at the NPL.

87

Fr

Francium

Francium, the last of the alkali metals, was discovered in 1939, by French physicist Marguerite Perey. The element is named after Perey's native country. Thirty-four isotopes are known, but even the most stable of them, francium-223, has a half-life of just 22 minutes. Nevertheless, francium does occur naturally, although probably not more than a few grams at any one time in the whole world: it is the product of the decay of other radioactive elements, most notably actinium (see page 95). The largest sample of francium ever prepared, consisting of only about 300,000 atoms, was made by bombarding gold atoms with oxygen atoms.

ATOMIC NUMBER: 87

ATOMIC RADIUS: 260

OXIDATION STATE: 1

ATOMIC WEIGHT: 223

MELTING POINT: 23°C (73°F), estimated

BOILING POINT: 680°C (1,250°F), estimated

DENSITY: 1.90 g/cm³

ELECTRON CONFIGURATION: (Rn) 7s¹

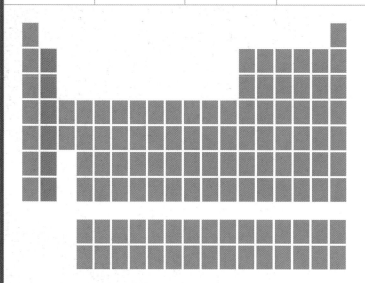

4

Be

Beryllium

12

Mg

Magnesium

20

Ca

Calcium

38

Sr

Strontium

56

Ba

Barium

88

Ra

Radium

The Alkaline Earth Metals

The elements of Group 2 are less reactive versions of the Group 1 elements. Like their more excitable cousins, these elements react with water and acids, producing hydrogen gas. But while Group 1 elements react explosively with cold water and even spontaneously with air, these Group 2 elements only react with water if it is very hot. The lesser reactivity is due to the electron configuration: the atoms of these elements have two electrons in their outer shell, compared with just one for the Group 1 elements.

Compared with its Group 1 equivalent, immediately to its left in the periodic table, each Group 2 element has an extra proton and an extra electron. The result is that the electrons are held more tightly and it takes more energy to remove them. However, once those two electrons are removed, the atoms have stable, filled outer shells, and the result is a doubly charged positive ion that can cling to negative ions to make very stable compounds (see page 11). For this reason, these elements are not found in nature in their pure state.

These elements are stronger, denser and better conductors of electricity than their Group 1 counterparts. In the Middle Ages, the term "earth" was applied to substances that do not decompose on heating, as is true of the oxides of calcium and magnesium in particular. The "alkaline" part of the group's name relates to the fact that the oxides of Group 2 elements all dissolve in water, albeit sparingly, producing alkaline solutions. However, the lightest of the Group 2 elements, beryllium, defies most of the properties described above, and is markedly different from the others.

4

Be

Beryllium

ATOMIC NUMBER: 4

ATOMIC RADIUS: 105 pm

OXIDATION STATES: +1, +2

ATOMIC WEIGHT: 9.01

MELTING POINT: 1,287°C (2,349°F)

BOILING POINT: 2,469°C (4,476°F)

DENSITY: 1.85 g/cm³

ELECTRON CONFIGURATION: [He] 2s²

Beryllium is named after the precious stone beryl. French chemist Nicholas-Louis Vauquelin discovered beryllium in a sample of beryl, in 1798 – although it was another 30 years before anyone could prepare a sample of the pure element. Despite its presence in precious stones and many other minerals, beryllium is actually very rare: it accounts for about two parts per million by weight of Earth's crust, and about one in every billion atoms in the Universe at large. Extraction of beryllium is a complicated process, the last stage of which involves heating beryllium fluoride (BeF_2) with another Group 2 element, magnesium. Only a few hundred tonnes of metallic beryllium is produced each year.

About two-thirds of the beryllium produced is used to make "beryllium copper" alloy, which contains up to 3 per cent beryllium by weight. This alloy is remarkably elastic and hard-wearing, and is used to make springs; it is also used to make tools for use in hazardous environments in which there are flammable gases, because it produces no spark when struck. Unusually for a metal, beryllium is rather transparent to X-rays, and beryllium compounds are used to make windows in X-ray tubes and detectors. Conversely, it is highly reflective of infrared light, and it can be worked very precisely to a polished finish, so it is used to make mirrors for orbiting infrared telescopes.

Beryllium is the odd one out among the alkaline earth metals: it does not form ions. As a result, beryllium compounds are all covalent, rather than ionic (see page 11).

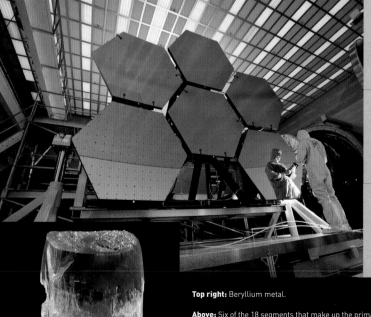

Top right: Beryllium metal.

Above: Six of the 18 segments that make up the primary mirror of the James Webb Space Telescope. The segments are made from beryllium, coated with a thin layer of gold.

Left: The precious gemstone beryl, composed chiefly of beryllium, aluminium, silicon and oxygen. Pure beryl is colourless but small amounts of other elements impart various colours.

12

Mg

Magnesium

ATOMIC NUMBER: 12

ATOMIC RADIUS: 150 pm

OXIDATION STATES: +1, +2

ATOMIC WEIGHT: 24.31

MELTING POINT: 650°C (1,201°F)

BOILING POINT: 1,091°C (1,994°F)

DENSITY: 1.74 g/cm³

ELECTRON CONFIGURATION: [Ne] 3s²

Magnesium is the eighth most abundant element in Earth's crust. It is found in the common rock-forming minerals olivine and pyroxene, and there is more than a million tonnes of it per cubic kilometre of seawater. The average adult human body contains about 25 grams of this essential element; more than 300 important biochemical reactions depend upon the presence of magnesium ions. Green vegetables are good dietary sources of magnesium, since atoms of the element lie at the heart of the green pigment chlorophyll. Other good sources include peas, beans, nuts, whole grains and potatoes.

Right: Magnesium metal.

Below: A ribbon of metallic magnesium, ignited by the flame of a Bunsen burner, burns vigorously in air, producing a bright white light.

Magnesium is familiar to many people as a thin ribbon of greyish metal that burns vigorously with a very bright white flame. Before the invention of electronic flashguns, photographers used disposable flash bulbs containing powdered magnesium that was ignited by an electric spark. Today, powdered magnesium is still used to create flashes for live shows. It is also found in fireworks, creating sparks and increasing a firework's overall brightness.

It was Scottish chemist Joseph Black who first identified magnesium as an element in 1755. Black was experimenting with magnesia alba (magnesium carbonate, $MgCO_3$), which was – and still is – used as an antacid to relieve heartburn. Magnesia alba, which means "white magnesia", is an ore that was found in the Magnesia region of Greece. English chemist Humphry Davy first prepared the element magnesium, in an amalgam with mercury, in 1808 – and French chemist Antoine Bussy isolated the first samples of magnesium metal proper 20 years later. By the middle of the nineteenth century, magnesium was being extracted on an industrial scale.

Today, China produces around 90 per cent of the world's magnesium. The main method of extracting the metal involves electrolysis of magnesium chloride ($MgCl_2$), which is first prepared from magnesium ores. Magnesium is strong but light, and although it burns extremely well as a powder, it is surprisingly fire-resistant in bulk, and it used safely in an ever-growing range of applications. World production more than trebled between 2000 and 2010, partly as a result of the need for lighter structural components in cars and aircraft and partly because of the falling cost of producing the metal. Around 10 per cent of magnesium is used in steel production, to remove sulfur from the iron ore.

One familiar magnesium compound found in most bathrooms is talcum powder, which is made from the softest known mineral, talc $(H_2Mg_3(SiO_3)_4)$. Epsom salt (hydrated magnesium sulfate, $MgSO_4.7H_2O$) is the main ingredient of bath salts; and gardeners also use it to combat magnesium deficiency in soil. It is also commonly used as a laxative. Milk of Magnesia (magnesium hydroxide, $Mg(OH)_2$) is another compound used as a laxative, and also as an antacid.

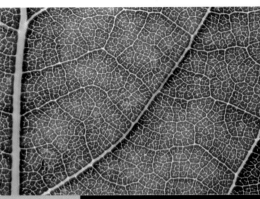

Above: Close-up photograph of an oak leaf, backlit to emphasize the green pigment chlorophyll. A magnesium ion sits at the centre of each chlorophyll molecule.

20
Ca
Calcium

ATOMIC NUMBER: 20
ATOMIC RADIUS: 180 pm
OXIDATION STATES: +1, +2
ATOMIC WEIGHT: 40.08
MELTING POINT: 842°C (1,548°F)
BOILING POINT: 1,484°C (2,703°F)
DENSITY: 1.55 g/cm³
ELECTRON CONFIGURATION: [Ar] 4s²

Above right: Pellets of dull, silver-grey calcium metal.

Like all the Group 2 (and Group 1) elements, calcium is too reactive to be found in its pure state in nature, despite the fact that it is the fifth most abundant element in Earth's crust. Only a few thousand tonnes of the pure element are produced each year.

Calcium was discovered in 1808 by Humphry Davy, who managed to separate it from a mixture of lime (CaO) and mercury(II) oxide (HgO). Davy named the element after the Latin word for lime, *calx*. Lime is the general term for any rock or mineral rich in calcium compounds.

There are four main calcium compounds found in rocks: calcium carbonate ($CaCO_3$), calcium sulfate ($CaSO_4$), calcium magnesium carbonate ($CaMg(CO_3)_2$) and calcium fluoride (CaF_2). Of these, calcium carbonate is the most widespread and the most interesting. There are several forms of calcium carbonate, each with a different arrangement of its atoms (crystal structure). The most important and most widespread of these is calcite, which is the main constituent of limestone and marble.

Limestone – and its more porous version, chalk – are sedimentary rocks that are made from the remains of countless tiny sea creatures, such as plankton. During their lifetime, those creatures absorbed carbon dioxide dissolved in the water and incorporated it into calcite to build their protective hard body parts. Marble is what is known as a metamorphic rock: it is limestone that has been altered by pressure and heat.

Since ancient times, people have been heating rocks containing calcium carbonate in lime kilns, driving off carbon dioxide to leave behind calcium oxide (CaO, also referred

to as lime; see previous page). Adding water produces calcium hydroxide ($Ca(OH)_2$), known as slaked lime, which ancient people used as a cement. The calcium hydroxide hardens, as it absorbs carbon dioxide from the air and forms calcium carbonate again. Today, that reaction is still central to the way cement works – although modern cement is more complex, involving calcium silicate.

Each year, more than a million tonnes of calcium carbonate are extracted from rocks for use in industry. The mineral has a wide range of uses besides the construction industry – including as a powder in hand cream, toothpaste and cosmetics, and in antacids. It is also used in the manufacture of food products, as a filler and white pigment in paper and paint, and in making glass and smelting iron ore.

Calcium is an essential element in nearly all living things. In humans, it is the most abundant metallic element; the average adult human body contains about 1.2 kilograms, of which around 99 per cent is tied up in calcium phosphate, the main constituent of bones and teeth. The rest is involved in vital functions, including cell division, healthy nerve and muscle function, the release of hormones and controlling blood pH.

Dairy products are well known as good dietary sources of calcium: for example, a single glass of milk contains about one-third of a gram. There is plenty of calcium in green vegetables, too, but much of it combines with oxalic acid also present to form a compound called calcium oxalate, which the body cannot absorb. Spinach has very high levels of calcium, but also of oxalic acid. Vitamin D is required for the proper uptake of calcium from the gut and is therefore needed for the growth and maintenance of bones; a deficiency of either calcium or vitamin D results in the disease rickets. Calcium compounds may be taken as dietary supplements or may be present in calcium-fortified foods; many calcium-rich foods are also fortified with vitamin D.

Above: Taj Mahal, in Agra, India, a mausoleum commissioned by Mughal emperor Shah Jahan in memory of his wife Mumtaz Mahal. The predominant building material is marble (calcium carbonate).

Below: False-colour scanning electron micrograph of a crystal of the ionic compound calcium phosphate (magnification approximately 250x). Bones and teeth are composed mostly of a variant of this mineral.

Below right: CT (computed tomography) scan of a human skull. Calcium accounts for approximately 100 grams of the 1 kilogram mass of an adult skull.

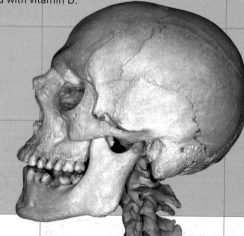

38

Sr

Strontium

ATOMIC NUMBER: 38
ATOMIC RADIUS: 200 pm
OXIDATION STATES: +1, +2
ATOMIC WEIGHT: 87.62
MELTING POINT: 777°C (1,431°F)
BOILING POINT: 1,382°C (2,520°F)
DENSITY: 2.64 g/cm³
ELECTRON CONFIGURATION: [Kr] 5s²

Above right: Silver-grey strontium metal.

Below: Fireworks display. Strontium compounds give firework explosions a red colour.

In 1790, Irish chemist Adair Crawford discovered a previously unknown mineral in a sample of rock that had been found in the Scottish town of Strontian. By studying its chemical properties, he surmised that it must contain a new element. In 1808, Humphry Davy isolated the element, which he named strontium, using electrolysis.

Until recently, one of the main uses of strontium was in televisions: strontium oxide (SrO) was added to the glass from which cathode ray tubes (CRTs) were made, to block X-rays produced inside the tubes. The rise of alternatives such as LCD televisions has all but halted production of CRTs. Strontium compounds are used in fireworks, normally to produce a bright red colour, and as phosphors in glow-in-the-dark toys.

Strontium has no biological role, but still the bones and teeth of everyone on Earth contain small amounts of it. This is because its chemical behaviour is very close to that of calcium, so it can easily replace calcium in the structural matrix of bones and teeth. The prescription drug strontium ranelate is used to increase bone mass and strength in patients with osteoporosis.

More than 30 isotopes of strontium are known; four are stable, and exist naturally, while the rest are unstable and radioactive. The radioactive isotope strontium-90 is produced in fission reactions, in nuclear reactors and nuclear weapons. Most of the strontium-90 in people around the world comes from the fallout from above-ground nuclear weapon tests. As a result of beta particles it emits as it decays (see page 9), strontium-90 can cause bone cancer and leukaemia, the likelihood increasing with dose. For most people, the presence of atoms of this unnatural isotope poses negligible risk. There is more strontium-90, however, in people born between 1945, the year of the first nuclear weapon test, and 1963, when a test ban treaty stopped most countries from testing above ground. The last above-ground nuclear weapon test took place in 1980. Strontium-90 levels were boosted somewhat – in part of the world at least – as a result of the Chernobyl nuclear reactor disaster in 1986.

<table>
<tr><td>56</td></tr>
<tr><td>Ba</td></tr>
<tr><td>Barium</td></tr>
</table>

ATOMIC NUMBER: 56
ATOMIC RADIUS: 215 pm
OXIDATION STATES: +2
ATOMIC WEIGHT: 137.30
MELTING POINT: 727°C (1,341°F)
BOILING POINT: 1,897°C (3,447°F)
DENSITY: 3.51 g/cm³
ELECTRON CONFIGURATION: [Xe] 6s²

Element 56, barium, is a dense grey metal. Its name derives from the Greek word *barys*, meaning "heavy". Like so many of the Group 1 and 2 elements, it was first isolated by Humphry Davy, in 1808. More than 30 years earlier, German chemist Wilhelm Scheele had realized that a previously unknown element was present in the mineral barite (barium sulfate, $BaSO_4$). This mineral was well known to chemists of the day, because after exposure to light, it glows in the dark.

Today, around six million tonnes of barite are produced each year. Barium metal is produced from this ore, normally by electrolysis of barium chloride ($BaCl_2$) produced from the barite. There are few uses for the metal itself; one application is that it is used to absorb traces of air in vacuum devices.

The most important compound of barium is barium sulfate, which is used to bulk out paints and plastics, to produce apple-green light in fireworks, and to increase the weight of drilling fluid that flows around drills used in oil wells. It is also the compound swallowed in a "barium meal", which a patient consumes before undergoing X-ray imaging to investigate digestive problems. Barium is opaque to X-rays. As a result, the patient's digestive tract appears in high contrast on an X-ray image, allowing radiographers and doctors to diagnose blockages, growths or problems with swallowing.

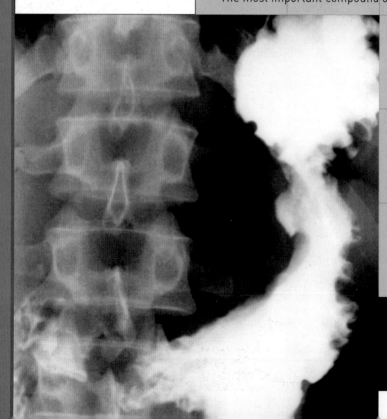

Top: Barium metal, kept in an inert atmosphere in a glass vial to prevent it reacting with oxygen in the air.

Left: X-ray of a human abdomen. The large solid white area is the stomach, which contains a "barium meal" of barium sulfate. The barium ions are opaque to X-rays.

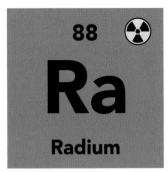

88 ☢

Ra

Radium

ATOMIC NUMBER: 88
ATOMIC RADIUS: 215 pm
OXIDATION STATES: +2
ATOMIC WEIGHT: 226
MELTING POINT: 700°C (1,292°F)
BOILING POINT: 1,737°C (3,159°F)
DENSITY: 5.5 g/cm^3
ELECTRON CONFIGURATION: [Rn] 7s^2

The element radium is highly radioactive – all of its isotopes are unstable. The element's name is derived from the Latin word for ray, *radius*. Radium is so unstable, in fact, that the minute quantities of the element existing naturally are all decay products from other radioactive elements.

Radium was discovered – and named – in 1898, by the Polish chemist and physicist Marie Curie and her husband, the French physicist Pierre Curie. The Curies had been studying the mysterious invisible rays produced by uranium, and had realized that the uranium ore they were working with was producing much more radiation than could be accounted for by uranium alone. Marie Curie and French chemist André Debierne were the first to extract pure radium, in 1910.

The beta rays produced by radium (see page 9) can cause certain other substances to glow – a phenomenon known as radioluminescence. This was the basis of the luminescent watch dials that were popular until the late 1930s: a small amount of radium was mixed with either zinc sulfide or phosphorus. After the toxicity of radium was discovered, safer radioactive sources replaced radium in luminescent paints.

In the 1910s and 1920s, before the dangers of radioactivity were fully known, radium was widely regarded as a wonder substance, and radium-containing items such as radium water, radium soap – and even radium wool to keep babies warm – were marketed as health products. Today, the only real applications of radium are in physics experiments.

Above right: Pitchblende (uraninite), the mineral in which radium was first discovered. Radium is a decay product of uranium, and is therefore found in tiny amounts in every uranium mineral.

Below: False-coloured tracks produced by alpha particles from a radium compound that had been left on the emulsion of a photographic plate.

Below right: Bulbs containing radon gas (1920s), used to produce radioactive "radon water", once thought to be healthful. The bulbs contained a small amount of radium, of which the radon gas was a decay product.

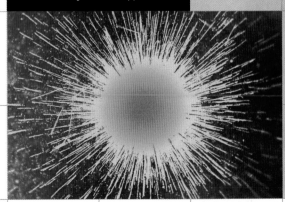

Interlude: The d-block and the transition metals

With just a glance at the periodic table, the middle section, containing Groups 3 to 12, stands out as being different. In particular, it has no elements in the first three periods – so it is not as tall as the block to its left (called the s-block) and the block to its right (the p-block). This misfit section, called the d-block, contains a host of versatile metallic elements with a wide variety of interesting compounds and uses. It includes some of the best known and most used metals, such as iron, gold, silver, copper and mercury.

This next section of the book, between pages 38 and 83, is devoted to the d-block elements, group by group. However, the elements with atomic numbers 104 to 112, which are part of the d-block, do not exist naturally, so they feature with the other transuranium elements, on pages 153–7. Further inspection of the periodic table highlights another irregularity: two elements in Group 3 are missing. In fact, they are part of a separate block of the periodic table: the f-block, which is explained on pages 84–5.

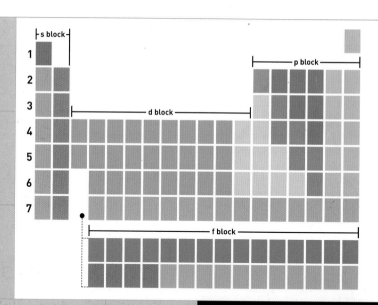

Understanding the d-block

There are two elements in Period 1 of the periodic table – hydrogen and helium – because there is only one s-orbital available for electrons at energy level n=1 (each orbital can hold up to two electrons; see page 8). In each of Periods 2 and 3, there is an s-orbital and three p-orbitals – space for eight electrons, which is why Periods 2 and 3 have eight elements each.

But in Period 4, in addition to the s- and p-orbitals, another type of orbital becomes available: the d-orbital. There are five d-orbitals at each energy level, making room for an extra 10 electrons. This is why Periods 4 and 5 contain 18 elements each. The appearance of f-orbitals in Period 6 complicates this still further (see page 9).

Above: The elements of Groups 1 and 2 have their one or two outermost electrons in an s-orbital – that is why there are only two columns at the left-hand end of the table. Those in the d-block have their outermost electrons in d-orbitals; as there are 5 d-orbitals, each with one or two electrons, this gives ten groups. The p-block contains elements in which the outermost electrons are in p-orbitals. There are three p-orbitals in each shell, producing a total of six columns (groups) in the p-block.

Opposite top: Shapes and relative orientations of the five p-orbitals.

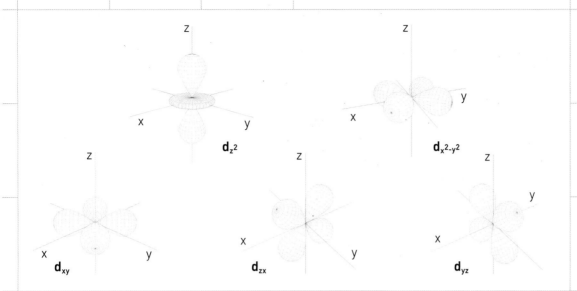

d_{z^2}

$d_{x^2-y^2}$

d_{xy}

d_{zx}

d_{yz}

The electrons in any given d-orbital actually have energies of the electrons in the shell below it. So in Period 4, the electrons fill a 4s orbital and 4p orbitals but the d-orbitals are 3d, at energy level n=3. This is like stacking shelves from the floor up, but leaving one empty, only to be filled when you are halfway through filling the next one up. The electron configurations of d-block elements are normally written in a way to reflect this oddity: the lower-energy d-orbital is shown last, since it is filled last. So, for example, compare the electron configuration of s-block element calcium with d-block element vanadium, shown immediately to the right:

Calcium	$1s^2$
	$2s^2\ 2p^6$
	$3s^2\ 3p^6$
	$4s^2$
Vanadium	$1s^2$
	$2s^2\ 2p^6$
	$3s^2\ 3p^6$
	$4s^2\ 3d^3$

The transition metals

Because the d-block acts as a bridge, or transition, between the s- and p-blocks, the d-block elements are often referred to as "transition metals". As defined by IUPAC a transition metal is "an element whose atom has an incomplete d subshell, or which can give rise to cations (positive ions) with an incomplete d subshell". This strict definition of what constitutes a transition metal rules out certain of the d-block elements. But in practice, the few elements that do not fit the definition are very similar to the "genuine" transition metals and they share the d-block, so they are normally included.

True to their name, the transition metals share properties common to all metals: they are shiny and silver-grey (exceptions being gold and copper), and they are good conductors of heat and electricity. They are malleable and ductile, meaning that they can be hammered or stretched, respectively, into different shapes – even mercury, a liquid at room temperature, is malleable and ductile when solid. Transition metals are generally much less reactive than the Group 1 and 2 metals; some, such as gold, are even found in nature in their pure state. Most transition metals form a range of brightly coloured compounds, such as the famously bright blue copper sulfate, the redness of rubies caused by the presence of chromium ions, and the reddish-brown of the iron oxide in rust. And many transition metals are used as catalysts – agents that speed up a reaction but do not actually take part in the reaction. Transition metal catalysts are commonplace in industry, and also in complex biological reactions.

Transition metals mix well with each other, and as a result, they form a huge range of alloys (an alloy is a mixture that contains at least one metal). Steel is an alloy of the transition metal iron (Fe) with other elements; stainless steel, for example, includes another transition metal, chromium. The pages of this section feature some of these alloys, each finely tuned for particular applications.

21

Sc

Scandium

ATOMIC NUMBER: 21
ATOMIC RADIUS: 160 pm
OXIDATION STATES: +1, +2, **+3**
ATOMIC WEIGHT: 44.96
MELTING POINT: 1,541°C (2,806°F)
BOILING POINT: 2,836°C (5,136°F)
DENSITY: 2.99 g/cm³
ELECTRON CONFIGURATION: [Ar] 3d¹ 4s²

Group 3 of the periodic table is made up of just two elements: scandium and yttrium. The two other spaces represent 16 elements each, comprising the 32 elements of the f-block. For more details, see pages 84–97.

Russian chemist Dmitri Mendeleev published the first periodic table in 1869 (see page 17). Mendeleev left several gaps, which he correctly supposed were as-yet undiscovered elements. Scandium was assigned to one of those gaps, after the element's oxide was discovered by Swedish chemist Lars Fredrik Nilson, in 1879. Since it was detected in a mineral found only in Scandinavia, Nilson called the new element scandium. It was not until 1937 that this elusive element was produced in its pure state.

Scandium is not particularly rare; its abundance in Earth's crust is about the same as that of lead, and greater than tin's. However, it is hardly ever found in high concentrations; it occurs instead as a trace element in over 800 different minerals. Only a few tonnes of scandium are produced each year, most of it used to make strong, lightweight alloys with aluminium. Soviet missiles launched from submarines had nose cones made of scandium-aluminium alloy – strong enough to penetrate through polar sea ice but light enough not to add much weight.

Above: Sample of pure scandium metal, a rare earth element but not a lanthanoid.

39

Y

Yttrium

ATOMIC NUMBER: 39
ATOMIC RADIUS: 180 pm
OXIDATION STATES: +1, +2, **+3**
ATOMIC WEIGHT: 88.91
MELTING POINT: 1,523°C (2,774°F)
BOILING POINT: 3,337°C (6,035°F)
DENSITY: 4.47 g/cm³
ELECTRON CONFIGURATION: [Kr] 4d¹ 5s²

Yttrium is often considered to be a rare earth element (see page 86), because its properties and applications are very similar to those elements, and it is most often found in the same mineral deposits as them. In fact, yttrium was the first rare earth element to be discovered, by Finnish chemist Johan Gadolin. In 1787, Gadolin received a sample of a newly discovered mineral from a quarry in the Swedish village of Ytterby, and in that sample he identified the oxide of the new element, in 1794. Yttrium was first isolated in its elemental state by German chemist Friedrich Wöhler, in 1828.

Above: Sample of the rare earth metal yttrium, a rare earth element but, as with scandium, not a lanthanoid.

Yttrium compounds have a variety of applications. Yttrium aluminium garnet (YAG) lasers are used in a wide variety of settings, including medical and dental procedures, surveying, cutting and digital communications. Yttrium oxide (Y_2O_3, yttria) added to zirconium oxide (ZrO_2, zirconia) gives yttria-stabilized zirconia, a very stable and inert ceramic with a number of applications, including uses in oxygen sensors, heat-resistant elements in jet engines and industrial abrasives and bearings.

22

Ti

Titanium

ATOMIC NUMBER: 22

ATOMIC RADIUS: 140 pm

OXIDATION STATES: +2, +3, **+4**

ATOMIC WEIGHT: 47.87

MELTING POINT: 1,665°C (3,029°F)

BOILING POINT: 3,287°C (5,949°F)

DENSITY: 4.54 g/cm³

ELECTRON CONFIGURATION: [Ar] 3d² 4s²

Group 4 of the periodic table includes three transition metal elements (see page 39) – titanium, zirconium and hafnium. It also includes the radioactive element rutherfordium, which is not found in nature and has an atomic number higher than uranium's. Rutherfordium is therefore featured in the section on the transuranium elements, on page 154.

In 1791, English amateur mineralogist Reverend William Gregor found some black sand near a stream. Gregor analysed the sand and found that it contained iron oxide and the oxide of another, unknown metal. Four years later, German chemist Martin Klaproth independently discovered the same unknown metal oxide in a mineral called rutile, and decided to call the metal "titanium", after the Titans of Greek mythology. It was not until 1910 that the metal was extracted in a very nearly pure form, and only since the 1950s has titanium been used in large quantities.

Titanium is the ninth most abundant element in Earth's crust, and the fourth most abundant metal. It is commonplace in igneous rocks and in black sands that are composed of igneous rocks. In the most common and widely used titanium mineral, ilmenite (iron(II) titanium(IV) oxide, $FeTiO_3$), titanium atoms are bonded to oxygen atoms and iron atoms. Titanium is extracted from ilmenite in an industrial process called the Kroll process, named after its creator, Luxembourgian chemist William Kroll. The ilmenite is heated with coke and chlorine gas, producing titanium(IV) chloride ($TiCl_4$); reacting this compound with magnesium metal removes the chlorine atoms, leaving pure titanium.

Because titanium is a reactive metal, oxygen from the air reacts with it rapidly and a thin layer of titanium oxide forms. This layer is responsible for the metal's apparent lack of reactivity and its resistance to erosion and corrosion. Titanium metal's inertness, strength and flexibility make it ideal for medical implants, such as artificial hip joints, artificial heart pumps and the cases of pacemakers. At high temperatures, the true reactivity of the titanium underlying the oxide layer becomes apparent. Titanium flecks or powder in fireworks burn vigorously once the fireworks have ignited, producing bright white light.

Titanium metal has the highest strength-to-weight ratio of all metals. It is as strong as steel but about half as dense. Around half of all the titanium metal produced is used in the aerospace industry – in particular, in high-performance alloys used to make fuselages and other structural components, turbine blades and tubes. Around 15 per cent of the weight of a Boeing 787 airliner is titanium, for example, and the metal is also an important constituent of

Top right: A bead of pure titanium metal.

Above: An artificial hip joint made from titanium, a light but strong metal. It consists of a shaft (bottom left) that is embedded in the femur (thigh bone) and a ball (right), which forms the joint with a socket (not seen) that is attached to the pelvis.

many satellites and the International Space Station. In addition to their specialized industrial and scientific applications, titanium metal and titanium alloys are also used in many consumer goods, such as golf clubs, jewellery, tennis rackets, watch straps and cars' alloy wheels. The shiny weatherproof coating of the Guggenheim Museum Bilbao, in Spain, is also made of a titanium-rich alloy. A particular alloy of titanium (with nickel) "remembers" its shape; this shape memory alloy is used in a variety of medical and industrial applications, and also in spectacle frames.

Only about 5 per cent of the titanium minerals mined each year are used to make titanium metal. Nearly all of the remaining 95 per cent is used to manufacture by far the most important of all titanium compounds: titanium(IV) oxide (TiO_2), also called titania. This bright white powder has many uses, and crops up in a wide range of familiar places. As a pigment, titanium dioxide is used in paints, cosmetics, toothpastes, tablets and foods such as sauces and cottage cheese. It is also used in many consumer plastics, which would otherwise appear dull grey. Titanium dioxide absorbs ultraviolet light, so it is a common ingredient in sunscreens.

When it absorbs ultraviolet light, titanium dioxide produces hydroxide ions (OH^-), which act as free radicals – reactive molecules that can break down oil, bacteria and other organic dirt particles. As a result, it is used in self-cleaning surfaces, including self-cleaning window glass and coatings for hospital wall and floor tiles that reduce rates of infection by killing bacteria.

Above: High-performance "blisk" (bladed disk) from a jet engine, cast from titanium.

Below: The Guggenheim Museum Bilbao, in Spain. The building is covered in more than 30,000 0.5-millimetre-thick pieces of titanium.

Below right: Rods of titanium – one of the most popular and convenient ways of supplying the metal.

40

Zr

Zirconium

ATOMIC NUMBER: 40
ATOMIC RADIUS: 155 pm
OXIDATION STATES: +1, +2, +3, **+4**
ATOMIC WEIGHT: 91.22
MELTING POINT: 1,854°C (3,369°F)
BOILING POINT: 4,406°C (7,963°F)
DENSITY: 6.51 g/cm³
ELECTRON CONFIGURATION: [Kr] $4d^2 5s^2$

Element 40, zirconium, has very similar properties, compounds and applications to titanium (see page 41). It, too, is a silver-grey transition metal that is intrinsically reactive but which forms a thin layer of oxide when exposed to the air, which renders it inert. And like titanium, zirconium is extracted using the Kroll process – in this case, by heating zirconium minerals with coke and chlorine. The resulting zirconium(IV) chloride is then heated with pure magnesium metal, which bonds with the chlorine atoms, leaving pure zirconium.

Zirconium was even discovered by the same person who discovered titanium, Martin Klaproth. In 1789, Klaproth was studying the mineral zircon, and managed to produce from it an oxide of an unknown metal. Since he had produced it from zircon, he called the oxide "zirconia", and the element zirconium. Swedish chemist Jöns Jacob Berzelius, a rival of Klaproth, was the first to produce nearly pure zirconium, in 1824.

Zircon mineral is mostly composed of zirconium silicate ($ZrSiO_4$). It had been known for hundreds of years, and was worn as a precious stone in Israel in Biblical times. The name "zircon" comes from an Arabic word, *zarqun*, which ultimately comes from an ancient Avestan (Indo-Iranian language) word, *zar*, meaning "gold". Zircons come in a variety of colours, depending upon the other elements present, but a common form is pale yellow-brown.

Pure zirconium is quite soft and very malleable and ductile. The addition of even small amounts of other elements makes it very hard, but it remains ductile. It also remains hard even at temperatures close to its very high melting point. As a result, zirconium alloys are used in situations where components can come under extreme heat and pressures, such as in nuclear reactor assemblies. Since zirconium has the unusual property that it does not absorb neutrons produced by nuclear fission reactions, it is also used in the cladding around nuclear fuel rods in some reactors.

Zirconia (zirconium(IV) oxide, ZrO_2) is a white compound similar to titania (titanium(IV) oxide, TiO_2). It is not as widely used as titania, but it has similar properties. Its hardness and its opacity to light means that it is used as an abrasive and as an "opacifier" in ceramics. One form of zirconia – with a crystal structure in which the atoms have a cubic formation – has an appearance and a hardness similar to diamond. This "cubic zirconia", often mistakenly called zircon, is used in jewellery as a cheaper alternative to real diamonds.

Above: Samples of cubic zirconia, a synthetic gemstone often used as an alternative to diamond. Cubic zirconia is made by melting naturally occurring zirconium oxide and then cooling it slowly. Colours are due to small amounts of other elements.

Top right: Pure zirconium, a silvery-white transition metal.

72

Hf

Hafnium

ATOMIC NUMBER: 72

ATOMIC RADIUS: 155 pm

OXIDATION STATES: +2, +3, **+4**

ATOMIC WEIGHT: 178.49

MELTING POINT: 2,233°C (4,051°F)

BOILING POINT: 4,600°C (8,312°F)

DENSITY: 13.3 g/cm³

ELECTRON CONFIGURATION: [Xe] $4f^{14}$ $5d^2$ $6s^2$

Element 72, hafnium, is another silver-grey transition metal, with properties, alloys and applications very similar to the other Group 4 metals titanium and zirconium (see pages 41 and 43). There are two particular uses of hafnium that stand out. First, like zirconium, hafnium is used in nuclear reactors; but unlike zirconium, hafnium does absorb neutrons produced in nuclear reactions. As a result, it is used in control rods; these slow nuclear reactions, by absorbing neutrons, when they are inserted into the reactor core. Secondly, in 2007, hafnium(IV) oxide found fame through its role in a new generation of microprocessor chips. The inclusion of hafnium oxide enabled the size of each transistor on the chips to be reduced. This has meant that the new chips can hold even more transistors than their predecessors, and are more energy-efficient. The hafnium oxide forms a "gate" that separates two electrical contacts of each transistor. Previously, silicon dioxide formed the gate, but shrinking the transistors any further would have caused electric current to leak across between the contacts.

Perhaps the most interesting thing about hafnium is the way in which it was discovered. The existence of "element 72" was predicted before the element was found. In 1913, English physicist Henry Moseley (see page 17) developed a technique that used X-rays to work out the amount of positive charge in the nuclei of atoms of a particular element. Moseley used his technique to test a theory Danish physicist Niels Bohr had proposed earlier that year. Bohr had suggested that electrons could only circle the atomic nucleus in certain orbits, determined by the amount of positive charge in the nucleus. According to Bohr, whenever an electron jumped down from one orbit to a lower one, the energy it lost would be emitted as a photon of electromagnetic radiation (see page 7) – visible light, ultraviolet light or even X-rays. Moseley measured the frequencies of X-rays given out by dozens of elements and verified Bohr's prediction. He also found that the nuclear charge increased by exactly the same amount from one element to the next – a fact that reinforced the then-new concept of atomic number and led to the discovery of the proton in 1917. As a result, Moseley was able to identify gaps in the list of elements, and one of those was an element with the atomic number 72. Hafnium was actually discovered in 1923, by Hungarian chemist George de Hevesy and Dutch physicist Dirk Coster. The element's name derives from *Hafnia*, the Latin name of Copenhagen, which was the birthplace of Niels Bohr and the city where the discovery took place.

Above right: Pure hafnium metal, a silvery-white transition metal.

Left: A wafer of nearly three hundred computer chips. Each chip has 1.4 billion transistors, each 22 nanometres across (22 billionths of a metre). At the heart of every transistor is a switch called a "gate", which is made with hafnium oxide.

23

V

Vanadium

ATOMIC NUMBER: 23
ATOMIC RADIUS: 135 pm
OXIDATION STATES: -1, +1, +2, +3, +4, **+5**
ATOMIC WEIGHT: 50.94
MELTING POINT: 1,910°C (3,470°F)
BOILING POINT: 3,407°C (6,165°F)
DENSITY: 6.10 g/cm³
ELECTRON CONFIGURATION: [Ar] 3d³ 4s²

Above right: Cuboid of pure vanadium.

Below: Crystal of the mineral vanadinite (lead chlorovanadate, $Pb_5(VO_4)_3Cl$). Vanadinite contains 11 per cent vanadium by mass, and is one of the main ores of the element.

Group 7 of the periodic table includes the elements vanadium, niobium and tantalum – all transition metals that are resistant to wear and corrosion. It also includes dubnium, an element that does not exist naturally and has an atomic number higher than uranium's – and therefore features in the section on the transuranium elements, on pages 153–157.

Vanadium was discovered in 1801, then undiscovered and later rediscovered – and it had several names before its modern one was chosen. It was Spanish mineralogist Andrés Manuel del Río who first suggested the existence of this new element, after studying a red-brown lead ore, known today as vanadinite, which is prized by mineral collectors to this day for its luscious red crystals. At first, del Río called the new element *Panchromium*, meaning "many colours", because of the variety of colourful compounds it produced. But by 1803, he was calling it *Eritrono* (or *erythronium*, from the Greek word for "red"), because a white powder produced from the mineral turned red whenever he heated it or reacted it with acids. In 1805, del Río gave a sample of the white powder to the renowned German explorer and scientist Alexander von Humboldt, who sent it on to French chemist Hippolyte-Victor Collet-Descotils. The French chemist mistook the new element for chromium, which had been discovered just a few years earlier, and he advised Humboldt that there was no "new" element. Humboldt informed del Río, who publicly renounced his claim.

In the 1820s, Swedish chemist Niels Sefström found an element that had properties similar to both chromium and uranium. Sefström gave it the name *Odinium*, after the Norse god Odin. Influential Swedish chemist Jöns Jacob Berzelius took over Sefström's research, calling the new element *Erian*; German chemist Friedrich Wöhler called it *Sefströmium*, before Sefström himself finally plumped for *Vanadium* – partly because there were no other elements beginning with the letter "V"! The name is derived from *Vanadis*, which means "lady of the Vanir tribe" and

refers to the Norse goddess Freyja. Vanadium proved very difficult for nineteenth-century chemists to isolate in its pure form; English chemist Henry Enfield Roscoe was the first to manage it, in 1867.

Despite its confusing early history, vanadium is a typical transition metal, forming colourful compounds and strong, useful alloys. Like many of the transition metals, vanadium forms a number of different ions; these are responsible for the colourful solutions that vanadium compounds form when dissolved in water. The ion V^{2+} gives purple solutions, V^{3+} green, VO^{2+} blue and VO_4^{2-} yellow.

Certain medieval swordsmiths forged famously hard blades from a steel that became known as Damascus steel. Modern research has shown that the steel owed its hardness and sharp edge to carbon nanotubes (see page 106) and the chance presence of small amounts of vanadium in the iron ore that was used. Today, vanadium is used to make corrosion-proof steel with high tensile strength (it can be pulled with great force without stretching or breaking). Vanadium steels are used in a wide range of applications, including ball bearings and springs. One of the most common vanadium alloys is chrome vanadium, which is used to make hard-wearing tools. One growing application for vanadium is in vanadium redox batteries, which can store large amounts of electrical energy to level out variations in the power output from renewable sources such as wind power.

Only about 7,000 tonnes of the pure metal are produced each year. Vanadium(V) oxide (V_2O_5) is the most important compound of vanadium; it is extracted from vanadium ores, and also from crude oil residues from the petroleum industry. Vanadium(V) oxide is used in producing pure vanadium metal, and in the production of ferrovanadium, the starting point for vanadium steels. It is also used as a catalyst in the production of sulfuric acid.

Vanadium is an essential element for humans, but no one knows exactly why. It seems to be a catalyst for certain metabolic reactions. Only tiny amounts are needed – and luckily, any excess is excreted. Vanadium is found in shellfish, mushrooms and liver.

Above right: Colourful solutions of vanadium compounds in water. From left to right, the dissolved vanadium atoms are short of five, four, three and two electrons – in other words, in the oxidation states +5, +4, +3, +2.

Right: Crayfish – a good dietary source of vanadium. The amount of vanadium they contain depends upon the concentration of the element in the seawater in which they live.

Far right: Drill bits made from strong, corrosion-proof vanadium-steel alloy.

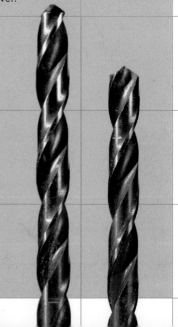

41

Nb

Niobium

ATOMIC NUMBER: 41
ATOMIC RADIUS: 145 pm
OXIDATION STATES: -1, +2, +3, +4, **+5**
ATOMIC WEIGHT: 92.91
MELTING POINT: 2,477°C (4,491°F)
BOILING POINT: 4,744°C (8,571°F)
DENSITY: 8.58 g/cm³
ELECTRON CONFIGURATION: [Kr] 4d⁴ 5s¹

The last two naturally occurring elements in Group 5 – niobium and tantalum – are very similar, and they are nearly always found together. It is no surprise, then, that their stories of discovery are suitably intertwined. Niobium was discovered first, in 1801, by English chemist Charles Hatchett, in a mineral that had been sent from a mine in Massachusetts. The mineral consisted of iron oxide and an unfamiliar substance. After carrying out chemical analyses, Hatchett concluded that the mystery substance was an oxide of a previously unknown metallic element: "one of those metallic substances which retain oxygen with great obstinacy". Having communicated with American scientists about the mineral, Hatchett decided to call the mineral "columbite" and the element "columbium", after the poetic name of the newly formed United States of America, Columbia.

A year later, Swedish chemist Anders Ekeberg discovered and named tantalum – but in 1809, influential English chemist William Hyde Wollaston proposed that tantalum and columbium were one and the same. It was not until 1844 that the confusion was resolved, when German chemist Heinrich Rose proved that columbite contained two distinct elements after all. In 1844 – just three years before Hatchett's death – Rose gave the non-tantalum element the name "niobium", after the Greek goddess Niobe, daughter of Tantalus. Despite Rose's definitive analysis and naming, chemists tended to use both columbium and niobium in equal measure until, in 1950, the International Union of Pure and Applied Chemistry decreed that element 41 should only be called niobium. In fact, some metallurgists and engineers (particularly in the USA) still use the alternative name and the symbol Cb.

The first nearly pure sample of niobium was produced in 1864. In its elemental form, niobium is a soft, silver-grey transition metal. Its crystal structure makes it particularly ductile and easy to work. Like many transition elements, niobium is added in tiny amounts during steel production – the main application for this element. Steels containing small amounts of niobium (less than 1 per cent) are hardwearing and strong, and niobium is

Below: Mining for coltan ore in the Democratic Republic of the Congo. Coltan is rich in the elements niobium (also known as columbium) and tantalum. Both elements are widely used in the modern electronics industry.

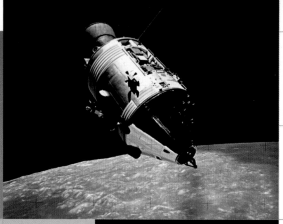

used in the production of structural steels. Larger amounts of niobium – typically around 5 per cent – are used in heat-resistant "superalloys" that are suitable for more specialized applications, such as heat shields, chemical industry crucibles and the nozzles of jet and rocket engines. The main nozzle of the Apollo Lunar Service Module was 89 per cent niobium.

At low temperatures, niobium is a superconductor: it offers no resistance to electric current. The same is true of certain niobium alloys, most notably niobium-tin and niobium-titanium. Wires made of these alloys are commonly used in superconducting electromagnets at the heart of medical scanners and particle accelerators, including the Large Hadron Collider at CERN.

Above left: Niobium alloy cable used as magnet windings in a particle accelerator at the DESY (Deutsches Elektronen Synchrotron) laboratory in Germany. The cable is superconductive (has no resistance to electric current) at very low temperatures.

Above right: Lunar Command Module of the Apollo 17 mission, 1972, in orbit around the Moon. The nozzle of the rocket engine (near the top of the picture) was made of a niobium alloy that was light but extremely heat-resistant.

73

Ta

Tantalum

ATOMIC NUMBER: 73

ATOMIC RADIUS: 145 pm

OXIDATION STATES: -1, +2, +3, +4, **+5**

ATOMIC WEIGHT: 180.95

MELTING POINT: 3,020°C (5,468°F)

BOILING POINT: 5,450°C (9,842°F)

DENSITY: 16.67g/cm³

ELECTRON CONFIGURATION: [Xe] $4f^{14}$ $5d^3$ $6s^2$

The name of element 73 is derived from the ancient Greek mythological character Tantalus, whose punishment for sacrificing his son was to stand in a lake in the afterworld, surrounded by water but unable to drink (his name is the source of the English word "tantalize"). Swedish chemist Anders Ekeberg chose the name after he discovered the element tantalum in 1802, because tantalum's oxide would not react with acids, even plentiful amounts of very strong acids, unlike the compounds of most other metals. The inert nature of tantalum makes it ideal for certain medical implants, including screws used to hold together bones, artificial joints and plates in the skull. Combined with its strength, this inertness makes it ideal in making vessels and pipes in the pharmaceutical and food industries. The pure elemental form of tantalum, first produced in 1903, is soft and ductile, like niobium. It is silver-grey, but with a bluish tinge. It has the fifth highest melting point and third highest boiling point of any element.

Tantalum metal owes its inertness to an extremely thin layer of tantalum(V) oxide (Ta_2O_5) that forms on its surface – the same reason why aluminium, titanium and zirconium are inert. While tantalum itself is a very good conductor of electricity, the oxide layer is a good insulator. This leads to the most important application of tantalum: the production of electronic components called capacitors. Inside a capacitor, two metal plates are separated by an insulating material – the thinner the better. An electric field passes through the insulator, known as a dielectric, so that changing electric currents in one part of a circuit can affect other parts, even though electric current cannot flow directly between them. Capacitors are widespread in almost every electronic circuit, and tantalum "pinhead" capacitors are of particular use in mobile phones and computers. The rapid rise of consumer electronics in the past few decades has led to an explosion in demand for tantalum.

The close association between tantalum and niobium (see page 47) means that the two elements are normally found together, and can be hard to separate. The main tantalum ore, tantalite, is often found in combination with the main niobium ore, columbite, and the combination is normally referred to as "coltan". The rising demand for tantalum – and therefore coltan – has become an international political issue. A 2003 United Nations report suggested that illegal smuggling of coltan out of the Democratic Republic of the Congo sustained military conflict and corruption, and that many consumer electronics companies were benefitting.

Above: Tiny "surface mount" capacitor, with compressed tantalum powder acting as one electrode and the natural tantalum oxide coating forming a crucial insulating layer.

Left: Tantalum metal discs.

Below: Image of a human retina, with a data diagram overlaid. A tumour, outlined in red, is to be treated with a proton beam. Fine, corrosion-resistant tantalum wires, to be placed at the points shown by small pink circles, will help to direct the beam and to conduct away the electric charge created.

24

Cr

Chromium

ATOMIC NUMBER: 24

ATOMIC RADIUS: 140 pm

OXIDATION STATES: -2, -1, +1, +2, **+3**, +4, +5, **+6**

ATOMIC WEIGHT: 52.00

MELTING POINT: 1,860°C (3,380°F)

BOILING POINT: 2,672°C (4,840°F)

DENSITY: 7.19 g/cm³

ELECTRON CONFIGURATION: [Ar] 3d⁵ 4s¹

Group 6 of the periodic table includes two well-known metals – chromium and tungsten – and one less well-known metal, molybdenum. It also includes seaborgium, a radioactive element that does not exist naturally, and features in the section on the transuranium elements, on pages 153–7.

Chromium – the twenty-first most abundant element in Earth's crust – was discovered in 1797, by Nicolas-Louis Vauquelin, who also discovered beryllium. Vauquelin was investigating a bright orange mineral called red lead, which had been discovered in a mine in Siberia in the 1760s. He carried out various chemical investigations on the mineral, producing a range of intriguing brightly coloured compounds as a result. He realized that the mineral contained a previously unknown element. Vauquelin named the new element after the Greek word *chroma*, meaning "colour", and he was able to isolate it in its shiny metallic form within a year.

The orange colour of the red lead Vauquelin investigated was due to the presence of chromate ions (CrO_4^{2-}). The colourful mineral was used to make chrome yellow, a popular pigment used by artists. During the twentieth century, chrome yellow was used in very large quantities in commercial paints – but it is no longer used because both lead and chromate ions are toxic. Chromate ions contain chromium in an oxidation state (see pages 12–13) of +6, or (VI); this means it forms bonds as if it were six electrons short. Another

Top: Highly lustrous nuggets of pure chromium.

Left: Precipitate of lead(II) chromate forming as a solution of potassium chromate is added to a solution of lead(II) nitrate. The yellow colour of the precipitate is characteristic of chromate ions (CrO_4)²⁻.

colourful chromium(VI) ion is the dichromate ion ($Cr_2O_7^{2-}$), which has been commonly used to tan leather since the nineteenth century, mostly in the form of potassium dichromate ($K_2Cr_2O_7$). Non-toxic chromium(III) compounds are also used to tan leather, and have largely taken over from their toxic chromium(VI) cousins. Not only is chromium in the oxidation state +3 non-toxic, it is essential to humans. Chromium(III) enhances the action of insulin and therefore aids the regulation of blood glucose. The human body needs only tiny amounts; oysters, egg yolk and nuts are good sources.

In 1837, scientists found they could deposit metallic nickel on to surfaces from solutions containing nickel ions. In 1848, French chemist Junot de Bussy managed to do the same with chromium. For the rest of the century, scientists tried to perfect chromium electroplating, to produce an attractive mirror-like surface and provide protection against corrosion. Commercial chrome plating finally became a reality in the late 1920s, using a solution containing chromate ions, but only became widespread after the Second World War, in particular on motor cars and aeroplanes. In the 1970s, scientists perfected the process of electroplating plastic surfaces – and since then, many decorative parts of cars' bodywork and other products have been made of chrome-plated plastics, which offer the sheen and protection of chrome on a light, cheap surface.

While chrome plating was slowly making its presence felt in the modern world, another even more widespread application of chromium had already taken hold: stainless steel. An alloy of iron with only tiny amounts of carbon but more than 10 per cent of chromium by mass, stainless steel was first produced commercially in the 1920s, and it is still the alloy of choice for cutlery, saucepans, many surgical instruments and architectural features. Stainless steel is cheap, can be produced in large amounts, and is resistant to rusting. Chromium is used in other alloys, including nichrome wire (nickel-chrome), for heating elements in toasters and hairdryers.

Top: Cloud Gate sculpture in Millennium Park, Chicago, USA. The sculpture is made from stainless steel, which contains more than 10 per cent of chromium by mass.

Above: A large emerald in a platinum pendant, surrounded by diamonds. Emerald is a form of the precious stone beryl. Pure beryl is colourless; the green colour is due to the present of chromium.

Left: 1965 Cadillac, with chrome-plated steel bumpers, hubcaps and trims.

42

Mo

Molybdenum

ATOMIC NUMBER: 42
ATOMIC RADIUS: 145 pm
OXIDATION STATES: -2, -1, +1, +2, +3, **+4**, +5, **+6**
ATOMIC WEIGHT: 95.94
MELTING POINT: 2,620°C (4,750°F)
BOILING POINT: 4,640°C (8,380°F)
DENSITY: 10.22 g/cm³
ELECTRON CONFIGURATION: [Kr] 4d⁵ 5s¹

The second element of Group 6, molybdenum, is another hard, grey transition metal in its elemental form. Like most transition metals, it is used in a variety of alloys, to which it gives strength and resistance to high temperatures.

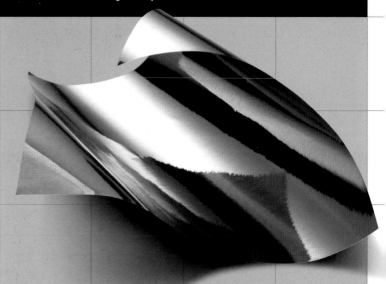

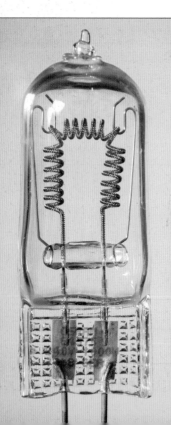

Molybdenum is fairly rare – about the 50th most abundant element in Earth's crust. It occurs in many different minerals, often associated with lead or calcium, but the main commercial source of the element is the mineral molybdenite. The dark grey colour of molybdenite made early mineralogists mistake it for lead ore or graphite; the old name for molybdenite was "molybdena", which derived from the ancient Greek word *molybdos*, meaning "lead". In 1778, Swedish-German chemist Carl Wilhelm Scheele proposed that a previously unknown element was present in molybdena, and three years later, another Swedish chemist, Peter Jacob Hjelm, isolated metallic molybdenum for the first time.

Molybdenum is used as a catalyst to speed up certain processes in the chemical industry. It is also a very important biological catalyst. Many enzymes (biological catalysts) contain molybdenum atoms – most importantly, the enzyme nitrogenase, found in certain bacteria, which "fixes" nitrogen from the air to form ammonia (NH₃). The ammonia is a raw material for building proteins (see page 117), essential compounds in all living things. Certain plants host nitrogen-fixing bacteria in their roots, and animals depend upon plants as their source of nitrogen. Some human enzymes also depend upon molybdenum, so humans require a small amount as part of their diet. Good sources of molybdenum include pulses such as lentils and beans, whole grains, and certain meats, including lamb and pork.

Top: A foil of lustrous elemental molybdenum.

Left: Halogen lamp used in studio lighting. The coiled filament is made of tungsten, but the connecting wires and supports are made of molybdenum, which makes a good seal with the glass bulb.

74

W

Tungsten

ATOMIC NUMBER: 74

ATOMIC RADIUS: 135 pm

OXIDATION STATES: -2, -1, +1, +2, +3, **+4**, +5, **+6**

ATOMIC WEIGHT: 183.84

MELTING POINT: 3,415°C (6,180°F)

BOILING POINT: 5,552°C (10,030°F)

DENSITY: 19.25 g/cm^3

ELECTRON CONFIGURATION: [Xe] 4f^{14} 5d^4 6s^2

Like the first two elements of Group 6, tungsten is a silver-grey transition metal. It has the highest melting point and boiling point of any metal – and second highest of all the elements, after carbon. Tungsten is considerably more dense than lead, and has almost exactly the same density as gold. There have been cases in which counterfeiters have produced bars of tungsten coated with gold and passed them off as gold bullion.

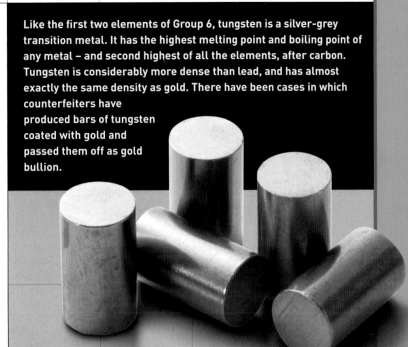

Above right: Samples of pure tungsten.

Below: Drill bit made from tungsten carbide, a compound consisting of equal numbers of tungsten and carbon atoms. Tungsten carbide is one of the hardest substances known, and has a very high melting point of 2,870 °C.

The name "tungsten" was originally used to describe the element's minerals; it comes from the Swedish for "heavy stone". Tungsten was recognized as an element in 1781, by Swedish-German chemist Carl Wilhelm Scheele, and was first isolated two years later. The symbol "W" relates to tungsten's alternative name, wolfram – from the German for "wolf spittle". Ores of tin and tungsten often occur together, and tungsten minerals produced a spittle-like foam during the smelting of tin. The foam reduced the yield of tin, and was thought to devour the tin as a wolf devours sheep. Most tungsten is derived from the ores scheelite and wolframite. China accounts for about three-quarters of the world's reserves and production.

More than half of all tungsten produced is used, in combination with carbon, to make the compound tungsten carbide (WC). This ultra-hard material was discovered accidentally in 1890, by French chemist Henri Moissan, as he was attempting to synthesize artificial diamonds. Tungsten carbide is used to make drill bits, circular saw blades and machine tools. Some cold-weather car and bicycle tyres have studs of tungsten carbide embedded into them, to improve traction in icy conditions. Tungsten carbide has also become popular as a material for making jewellery – particularly rings. And, since 1957, it has been used to make the ball in most ballpoint pens.

There are a few other important compounds of tungsten – most notably tungsten trioxide, which is used in fireproof fabrics, in "smart" windows (which can be darkened with the flick of a switch) and as a yellow pigment. Other tungsten compounds used as pigments include zinc tungstate and barium tungstate, both brilliant white. Some compounds of tungsten that also contain phosphorus and molybdenum are used as dyes and pigments for printing inks, plastics and rubber. As with most transition metals, some of tungsten's compounds are also used as catalysts in oil refining and the chemical industry.

There are many applications in which tungsten is used in its elemental form. Its high melting point and hardness make it ideal for use in the nozzles of rocket engines and as the electrode in very high-temperature gas tungsten arc welding. Tungsten metal is used to make the filaments of incandescent lamps and X-ray generators. Although incandescent lamps have largely been superseded by energy-saving fluorescent ones, tungsten filaments are still used in halogen lamps. Many car windscreen de-misters use tungsten heating elements embedded in the glass.

Tungsten is also used, mixed with other elements, in many high-performance alloys. The most important are the "high-speed steels", which have at least 7 per cent tungsten, vanadium and molybdenum. High-speed steels are used to make machine tools, and also in high-temperature applications such as the valve outlets of engines. Other tungsten alloys are used to make armour-piercing weapons, replacing the less desirable alternative: depleted uranium. Similarly, while ordinary bullets are still normally made using lead, tungsten alloys are becoming more common, as regulations about releasing lead into the environment become more strict. The high density and relatively low cost of another heavy tungsten alloy makes it ideal as the tiny, off-centre weight that spins around to make mobile phones vibrate.

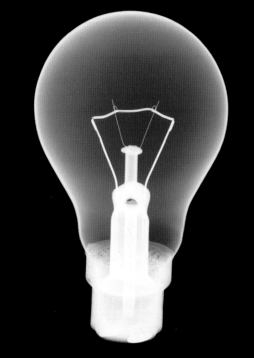

Above: X-ray image of an incandescent lamp, with a tightly coiled tungsten filament that glows when current flows through it. Today, energy-saving fluorescent lamps, which do not have a filament, have largely superseded incandescent lamps.

Left: NASA's Ares 1-X test rocket lifts off from launch pad 39B at the Kennedy Space Center in Cape Canaveral, Florida on 28 October 2009. Tungsten's hardness and high melting point makes it ideal for rocket nozzles and nose cones.

25

Mn

Manganese

ATOMIC NUMBER: 25

ATOMIC RADIUS: 140 pm

OXIDATION STATES: -3, -2, -1, +1, **+2**, +3, **+4**, +5, +6, **+7**

ATOMIC WEIGHT: 54.94

MELTING POINT: 1,245°C (2,273°F)

BOILING POINT: 1,962°C (3,563°F)

DENSITY: 7.47 g/cm^3

ELECTRON CONFIGURATION: [Ar] 3d^5 4s^2

Group 7 of the periodic table includes the elements manganese, technetium and rhenium. It also includes bohrium, an element that does not exist naturally and has an atomic number higher than uranium's – and therefore features in the section on the transuranium elements, on pages 153–7.

Above right: Nugget of elemental manganese.

Below: An alkaline battery, cutaway to reveal the black manganese dioxide powder that forms the cathode (positive terminal).

Manganese is the third most abundant transition metal in Earth's crust, and the twelfth most abundant element overall. Like many transition metals, manganese is essential to all living things, although the amounts required are tiny. Many different enzymes rely upon manganese atoms for their proper functioning, including the "oxygen-evolving complex" that releases oxygen from water as part of photosynthesis in plants. In humans, manganese is involved in a wide range of processes, including the formation of connective tissue, production of blood clotting factors and regulation of metabolism. An adult human body contains around 12 milligrams of manganese, and around 5 milligrams are required each day from a person's diet; foods rich in the element include eggs, nuts, olive oil, green beans and oysters. In larger amounts, however, manganese is toxic; long-term daily uptake of 10 milligrams or more can cause a range of problems, including neurological disorders, and hinder the uptake of dietary iron.

The existence of manganese as an element was first proposed by Swedish-German chemist Carl Wilhelm Scheele, in 1774. Four years earlier, Austrian chemistry student Ignatius Kaim had already isolated a small amount of the brittle metal from a manganese(IV) oxide – but he hadn't realized that it was an unknown element. Swedish chemist Johann Gottlieb Gahn is generally credited with being the first to isolate manganese, after he produced the metal in 1774.

The investigations into manganese had been the result of experiments carried out into a mineral containing manganese(IV) oxide (MnO_2). Glassmakers in the Roman Empire and even as far back as Ancient Egypt used this mineral to make their glass colourless, and in some cases to give the glass a pale purple hue. From the sixteenth century, glassmakers called this black mineral *magnesia negra*, because there were deposits of it in Magnesia, in Greece. This fact was the stimulus for German scholar Philipp Buttmann to name the element "manganese" in 1808 – the same year English chemist Humphry Davy discovered magnesium (see page 32), another element derived from a mineral found in Magnesia.

Manganese(IV) oxide is still used as a pigment in glass and ceramics today, but its main use is in dry cell batteries. Mixed with carbon powder to increase the electrical conductivity, manganese ions accept electrons produced in the main reaction inside the battery, becoming "reduced" from oxidation state +4 to +3, so that the manganese(IV) oxide ends up as manganese(III) oxide (Mn_2O_3). Other important manganese compounds include the deep purple compound potassium manganate(VII), also known as potassium permanganate ($KMnO_4$), which has many uses, including as a disinfectant and an antiseptic.

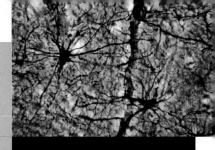

Manganese played an important role in the development of the modern steel industry. Already in the 1810s, people had discovered that adding small amounts of the element improved steel's strength and flexibility. Then, in the 1850s, when English industrialist Henry Bessemer was having trouble perfecting his process for producing large quantities of cheap steel, manganese solved the problem. English metallurgist Robert Mushet found that adding an ore rich in manganese would remove sulfur and oxygen, impurities that made the steel brittle and difficult to work. Mushet's discovery made Bessemer's process work. Today, more than 90 per cent of all the manganese produced is used in the steel industry – and modern steels nearly all contain manganese.

Above: Micrograph of astrocytes (support neurones) from a human brain. Manganese accumulates in these cells and, because of the element's neurotoxicity, it has been implicated as a cause of various neurodegenerative diseases.

43 ☢

Tc

Technetium

ATOMIC NUMBER: 43
ATOMIC RADIUS: 135 pm
OXIDATION STATES: -3, -1, +1, +2, +3, **+4**, +5, +6, **+7**
ATOMIC WEIGHT: (98)
MELTING POINT: 2,170°C (3,940°F)
BOILING POINT: 4,265°C (7,710°F)
DENSITY: 11.50 g/cm^3
ELECTRON CONFIGURATION: [Kr] $4d^5\,5s^2$

The name of this element is derived from the Greek word *tekhnitós*, which means "artificial"; technetium was the first element to be discovered only after being produced artificially. Only very tiny amounts of technetium occur naturally; every isotope of this element is unstable. Although two of the isotopes have half-lives of more than a million years, any technetium made inside the stars and supernovas that gave rise to the Solar System billions of years ago have long since decayed. The small amounts of technetium that do exist on Earth are the result of the disintegration of uranium nuclei in uranium ore, but they are extremely hard to detect.

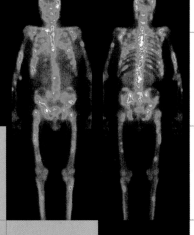

It is no surprise, then, that technetium was one of the elements that had not yet been identified when Dmitri Mendeleev drew up the first periodic table of elements in 1869 (see page 17). However, Mendeleev noted the gap in his chart below manganese and proposed the existence of an element that he called *ekamanganese* – and he correctly predicted many of its properties. The missing element was misidentified many times in the following decades. It was finally discovered in 1937, by Italian mineralogist Carlo Perrier and Italian-American physicist Emilio Segrè, after it had been made artificially in an early particle accelerator. Inside the accelerator, molybdenum foil was bombarded with deuterons (a type of particle composed of one proton and one neutron). Molybdenum has one less proton in its nucleus than technetium does, and the extra protons had transmuted some of the molybdenum into technetium.

Hundreds of kilograms of technetium are produced artificially each year – all of it radioactive, and most of it as a waste product of nuclear reactors. In its bulk elemental form, it is a shiny grey metal like most other d-block elements. One

Above: A false-colour bone scan of a human male, produced by detecting gamma rays emitted by the radioisotope technetium-99m injected into the patient. The "hot spots" indicate cancerous tissue.

relatively stable isotope, technetium-99, has a half-life of 211,000 years and decays by beta decay (see pages 9–10). One excited, metastable state of technetium-99 – called technetium-99m – is used in medicine as a radioactive tracer injected into a patient's bloodstream. The excited nuclei lose energy spontaneously, with a half-life of about six hours, each one emitting a photon of gamma radiation as it does so. The gamma radiation is detected and used to produce images of blood flow, organ function or cancerous areas.

Right: The first-ever generator of technetium-99m labelled radiopharmaceuticals, developed in 1958 by researchers at the Brookhaven National Laboratory, USA. Since then, hundreds of millions of scans have been carried out using technetium-99m.

Below: Small sample of the element rhenium.

While technetium was the first element to be discovered after being made artificially, rhenium was the last element to be discovered that has at least one stable isotope, and which therefore occurs in minerals. In its pure state, rhenium is more dense than gold, with a very high melting point.

75
Re
Rhenium

ATOMIC NUMBER: 75
ATOMIC RADIUS: 135 pm
OXIDATION STATES: -3, -1, +1, +2, +3, **+4**, +5, +6, +7
ATOMIC WEIGHT: 186.21
MELTING POINT: 3,182°C (5,760°F)
BOILING POINT: 5,592°C (10,100°F)
DENSITY: 21.02 g/cm³
ELECTRON CONFIGURATION: [Xe] $4f^{14} 5d^5 6s^2$

Like technetium, rhenium fills a gap identified by Mendeleev in his periodic table. Just as he had called the element one below manganese (technetium) *ekamanganese* (see opposite), he called the missing element two below *dwimanganese* – *eka* is a Sanskrit word meaning "one", *dwi* meaning "two". Rhenium was discovered in 1925, using X-ray spectroscopy (see page 44) of platinum minerals, by the German chemists Walter Noddack, Ida Tacke (who later married Noddack) and Otto Berg. The three scientists named the element after *Rhenus*, the Latin word for the major river of their homeland, the Rhine. Some scientists disputed the discovery, but in 1928, Noddack, Tacke and Berg managed to produce more than a gram of the new element, from nearly 700 kilograms of a molybdenum ore.

Rhenium does exist in nature, but in minuscule amounts; it is the fourth from last most abundant element in Earth's crust. For such a rare element, rhenium has a surprising range of applications – and around 50 tonnes of it are produced each year, mostly from the ore molybdenite. When the ore is roasted during the production of molybdenum, rhenium combines with oxygen to form rhenium(VII) oxide (Re_2O_7), which escapes as a gas. The gas is absorbed as it escapes in the flue, and processed to extract the rhenium.

The biggest application of rhenium is in alloys that are expected to withstand high temperatures – and most rhenium-containing alloys are used in jet engines and other gas turbines. Another significant application of rhenium is as a catalyst in the oil industry, to speed up the cracking of large hydrocarbon molecules in oil or gas. Two radioactive isotopes of rhenium are used in radiotherapy (known in the USA as radiation therapy) to treat liver, prostate and bone cancers, as well as reducing the pain associated with certain cancers.

Even though rhenium is rare, it is found in certain iron ores, and geologists use a method based on the decay of a radioactive rhenium isotope to date rocks more than a billion years old. Rhenium-osmium dating involves radioactive isotope rhenium-187, which decays to osmium-187 with a half-life of more than 40 billion years.

26

Fe

Iron

ATOMIC NUMBER: 26

ATOMIC RADIUS: 140 pm

OXIDATION STATES: -2, -1, +1, **+2**, **+3**, +4, +5, +6, +7, +8

ATOMIC WEIGHT: 55.84

MELTING POINT: 1,535°C (2,795°F)

BOILING POINT: 2,750°C (4,980°F)

DENSITY: 7.87 g/cm³

ELECTRON CONFIGURATION: [Ar] 3d⁶ 4s²

The most important element of Group 8 of the periodic table is iron, but this group also includes the little-known elements ruthenium and osmium. It also includes the radioactive element hassium, which does not exist naturally and has an atomic number higher than uranium's, and therefore features in the section on the transuranium elements, on pages 153–7.

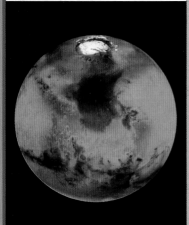

Iron is one of the most familiar of all the elements – in particular, when it is alloyed with carbon and other elements in the steel objects that surround us in everyday life. It was known to ancient metalworkers, and is one of the seven "metals of antiquity" (see page 76); in Ancient Greek and Roman mythology, iron was associated with the planet Mars, probably because of its association with weapons of war, Mars being the god of war. It is perhaps a pleasing (or even "ironic") coincidence that the red colour of planet Mars is due to the rocks and dust of red iron(III) oxide (Fe_2O_3) across the whole planet's surface.

The Latin word for iron is *ferrum*, which is why the chemical symbol for iron is Fe. As a pure metal, iron is soft, shiny, silver-grey, and malleable and ductile, like most other transition metals. Together with nickel and cobalt, iron is one of just three ferromagnetic elements: it can be magnetized, and is attracted to a magnet. Iron is by far the most magnetic, and even many iron ores can be magnetized. The phenomenon of magnetism was discovered in naturally magnetized pieces of an iron ore called magnetite. These pieces became known as lodestones, meaning "leading stones", because if they are free to turn, they always line up north–south, and could be used as compasses to lead the way.

Iron is a fairly reactive metal, readily combining with oxygen and water to form a variety of iron oxides and hydroxides. One of the most familiar products of these reactions is red-brown hydrated iron oxide, commonly known as rust. Mixing iron with various other elements can prevent it from rusting and change its properties in a host of other ways. Iron is by far the most common element used to make alloys, partly because it is plentiful, easy to extract and cheap.

Iron was not always cheap, however. In early civilizations, iron was more highly prized than gold and silver. Before the

Top right: Pure iron, which is bright, silvery and lustrous.

Above: Telescopic view of the planet Mars, whose red colour is due to the presence of iron oxides in its surface minerals. The planet appears red even to the naked eye, although it appears as no more than a point of light.

Right: Powerful electromagnetic crane moving ferrous (iron-containing) scrap.

dawn of the Iron Age, metal workers could only work with iron that had literally fallen from the sky – around one in 20 meteorites is made predominantly of iron, with a small amount of nickel. But in the Middle East in the third millennium BCE, people began obtaining iron from its ores by smelting. The most common iron ores are magnetite (Fe_3O_4) and haematite (Fe_2O_3), and they occur across the world. It is not known who first smelted iron ore to obtain the metal, but by around 1000 BCE, smelting was fairly common in several parts of the Middle East, Asia and Africa, spreading to Europe over the next few hundred years. Iron was still scarce and highly prized, because only small amounts could be made. In Homer's epic poem *Iliad*, Achilles organizes a competition to see who can hurl a lump of iron the furthest; the reward is a plentiful supply of iron.

Early iron smelting took place in furnaces called bloomeries. Inside, charcoal was burned, and attained a high temperature as a result of bellows pumping air into the fire. Iron ores are mostly iron oxide, and carbon monoxide (CO) from the burning charcoal dragged oxygen away from the ore, leaving metallic iron. The resulting drops of iron fell to the bottom, forming a porous mass called a bloom, which also contained slag – a combination of the other elements in the ore, such as silicon and oxygen. Blooms had to be heated and hammered – or wrought – to obtain fairly pure iron.

In Ancient China, metal workers smelted iron ore in a different type of furnace: the blast furnace. In Europe, the blast furnace was invented independently in the fifteenth century. In a blast furnace, limestone is added to the mix of iron ore and charcoal. The limestone helps to remove impurities more effectively, while a continuous blast of air, originally provided by water-powered bellows, enables a hotter furnace overall. The iron and the slag are both liquid, and can be separated more easily than in a bloomery. Blast furnaces run continuously for months or even years at a time, while bloomeries worked in small batches. The use of charcoal in blast furnaces led to widespread deforestation across Europe, and was a limiting factor in the production of iron. In 1709, English iron master Abraham Darby introduced coke, roasted coal, as an alternative fuel – an enabling step in the Industrial Revolution.

Above: Rusted iron chain links. Rust is a mixture of various iron oxides and hydroxides, and is weaker and more brittle than iron.

Left: Spikes in a ferrofluid – a suspension of tiny iron particles in a thick liquid. The spikes are due to the presence of a magnetic field.

Below left: Meteorite composed almost exclusively of iron and nickel. So-called iron meteorites were the first source of iron available to ancient metallurgists.

The liquid iron produced in blast furnaces can be "cast" in moulds. However, because of the higher temperature inside a blast furnace, the iron absorbs carbon; the resulting "pig iron" contains as much as 4 per cent carbon, making it too brittle to be workable. Therefore, pig iron has to be refined to remove the carbon. Originally, this was a laborious task involving re-melting the iron to produce a bloom, and then hammering the bloom to produce wrought iron as before. But in the 1780s, English inventor Henry Cort introduced a technique known as puddling. Inside a puddling furnace, hot air from a furnace re-melted the iron, and oxygen from the air combined with carbon in the pig iron, leaving nearly pure iron. A skilled worker stirred molten pig iron, gathering the pure iron as a "puddle" that was then removed.

Unlike cast iron, wrought iron is malleable; however, it is too soft for many applications, and

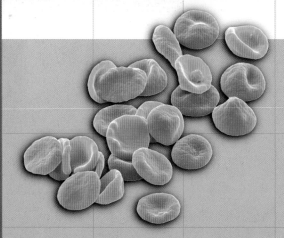

unlike cast iron, it rusts. The answer is to make steel – with more carbon than wrought iron and less than cast iron. Steel is hard, workable and less prone to corrosion. To produce it, the nearly pure wrought iron produced in puddling furnaces had to be processed yet again, to add carbon. In the 1850s, English metallurgist Henry Bessemer invented a process that could convert carbon-rich pig iron directly into steel. A Bessemer converter was a huge, pear-shaped receptacle containing the molten pig iron, through which compressed air was blown. The oxygen from the air removed just enough of the carbon, and also caused the converter to stay hot. A Bessemer converter could convert tonnes of pig iron directly to steel within half an hour. Modern steelmaking normally involves an updated version of Bessemer's process, using a lance that pumps pure oxygen, rather than air, through the pig iron. Detailed knowledge of the chemistry of iron has led to a vast range of different steels and other iron-based alloys.

Iron is not only important as a metal for making structures, vehicles and implements, it is also a vital element in all living things. Deep inside each cell, many vital reactions rely upon iron – including the synthesis of DNA and, in plants, photosynthesis. An adult human body typically contains about 4 grams of iron – more than half of it bound up in haemoglobin, the protein in red blood cells that is responsible for transporting oxygen in the bloodstream. Recommended dietary intake of iron is between about 10 and 25

milligrams per day, depending upon age and gender. A deficiency of iron limits the amount of oxygen delivered to the tissues, and leads to fatigue and lowered immunity – a condition called iron-deficiency anaemia – while iron in excess is toxic; the body cannot excrete iron and, left untreated, "iron overload" can be fatal.

It is not surprising that iron is so important to living things, because it is extremely plentiful. Iron accounts for around one-third of the mass of planet Earth as a whole – the largest proportion of any element, followed closely by oxygen. In Earth's crust, however, iron is only fourth most abundant by mass, after oxygen, silicon and aluminium. When Earth first formed, its composition was uniform throughout. Gravity pulled the young planet into a tight ball, and that compression generated enormous amounts of heat. The constant bombardment by asteroids and other objects in the young Solar System and the radioactive decay of unstable elements also generated heat. As a result, the planet's temperature rose above the melting point of iron. The molten iron, together with molten nickel, flowed down towards the centre, slowly at first. Friction generated more heat, melting more iron and resulting in a dramatic descent of iron and nickel that geologists call the "iron catastrophe". The molten elements displaced lighter elements, and the result was the layered structure of our planet today – iron-rich core at the centre, surrounded by the mantle and then the crust outermost.

Iron makes up nearly 90 per cent of our planet's core (with nickel accounting for the rest) but only about 6 per cent of the crust. The inner core is solid, and it spins within the liquid outer core. This rotation is the source of Earth's magnetic field, which stretches way out into space, creating a force field called the magnetosphere. This field

Top: Coloured scanning electron micrograph of human red blood cells, whose red colour is due to the iron-containing protein haemoglobin, which is responsible for carrying oxygen in the blood.

Right: The Iron Bridge, in Coalbrookdale, UK – the world's first cast-iron bridge, built in 1779, after improvements in iron production caused dramatic reductions in the metal's price.

diverts the solar wind – a stream of ionized particles speeding out from the Sun, which would otherwise cause untold problems for life on Earth, not least blowing most of the atmosphere and oceans off into space.

Also out in space, iron holds a special place in the story of the origin of the elements. The elements are all produced inside stars, by nuclear fusion. During most of their lifetime, stars produce helium, from the hydrogen raw material. But at the end of their life, stars fuse helium nuclei together to form nuclei of elements up to carbon and oxygen. In most stars, the process stops there, and the star cools uneventfully to form a white dwarf. But in really massive stars, the process continues up to iron. At the point when iron nuclei are made, these enormous stars suddenly collapse in on themselves, generating yet more heat – enough to fuse all the heavier elements and to create a colossal explosion, called a supernova, that sends most of the contents of the star far out into space. It is those elements that form new planetary systems, like our own Solar System.

Right: The 324-metre-tall Eiffel Tower, in Paris, France. Constructed in 1889, it is made from a latticework of wrought iron.

44
Ru
Ruthenium

ATOMIC NUMBER: 44
ATOMIC RADIUS: 130 pm
OXIDATION STATES: -2, +1, +2, **+3**, **+4**, +5, +6, +7, +8
ATOMIC WEIGHT: 101.07
MELTING POINT: 2,335°C (4,230°F)
BOILING POINT: 4,150°C (7,500°F)
DENSITY: 12.36 g/cm³
ELECTRON CONFIGURATION: [Kr] $4d^7 5s^1$

The lustrous, silvery-grey transition metal ruthenium is one of the least abundant of the naturally occurring elements – though not as rare as its relation directly below it in Group 8, osmium (see page 62). In nature, these two elements often occur together, with four other very rare elements that are neighbours in the periodic table. Together, the six related elements go by the name of the platinum group – see page 70.

While the other platinum group elements had been differentiated from platinum in the first few years of the 1800s, ruthenium was discovered much later. Russian chemist Karl Klaus first identified and isolated ruthenium as an element, in a sample of platinum ore, in 1844. He based the name on the word *Ruthenia*, a pseudo-Latin word for an ancient region of what is now western Russia and Eastern Europe.

Like all platinum group metals, ruthenium is generally very unreactive. It sometimes occurs in nature in its elemental state, uncombined with other elements, although normally it is mixed in with the other platinum group elements, requiring a multi-step procedure to separate it. Each year, about 20 tonnes of this rare metal are produced. Adding ruthenium to platinum or palladium results in very wear-resistant alloys that are commonly used to make high-performance electrical contacts – such as those in spark plugs – and to make fountain pen nibs. Ruthenium is sometimes used as an alloying element in platinum jewellery, too. Extremely heat-resistant superalloys involving ruthenium are used in jet

Above right: A nugget of pure ruthenium metal.

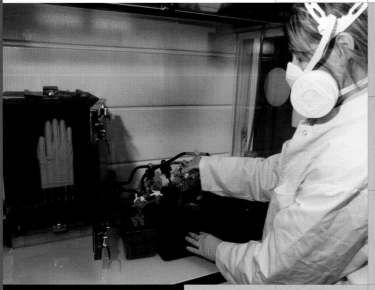

engines. Ruthenium and some ruthenium compounds make excellent catalysts, which are used in the pharmaceutical industry, for example.

Ruthenium has a growing number of applications in the electronics industry. Thin layers of the ceramic material ruthenium(IV) oxide (RuO_2) have been used in tiny chip resistors on circuit boards. A very thin layer of ruthenium is applied to the writable surface of some hard disks, greatly increasing the density of data that can be held there. Ruthenium red is a dye used to stain certain samples for viewing in an electron microscope – and a light-sensitive ruthenium dye is used in experimental dye-sensitized thin film solar cells that are robust and flexible, but low-cost. Ruthenium has no known biological role. However, ruthenium compounds will severely stain skin – and in even relatively small amounts, most ruthenium compounds are extremely toxic, as well as carcinogenic.

Above: Forensic scientist using ruthenium tetroxide vapour to detect fingerprints on a latex glove. The vapour reacts with fats in the fingerprints, staining them dark brown with ruthenium dioxide.

Below right: Elemental osmium, a bright, silvery metal.

76

Os

Osmium

ATOMIC NUMBER: 76

ATOMIC RADIUS: 130 pm

OXIDATION STATES: -2, +1, +2, +3, **+4**, +5, +6, +7, +8

ATOMIC WEIGHT: 190.23

MELTING POINT: 3,030°C (5,485°F)

BOILING POINT: 5,020°C (9,070°F)

DENSITY: 22.58 g/cm³

ELECTRON CONFIGURATION: [Xe] $4f^{14}\ 5d^6\ 6s^2$

The heaviest of the naturally occurring Group 8 elements is osmium. Among the rarest of all the stable elements, osmium makes up just 1 part per billion of Earth's crust by mass. In its pure form, osmium has the highest measured density of all elements – although calculations of what the density should be put it second, very slightly below iridium (see page 66). A block of osmium (or iridium) will weigh twice as much as a block of lead of the same volume.

Osmium is found in naturally occurring alloys of iridium and osmium – called either osmiridium or iridiosmium, depending upon which element dominates – as well as in platinum-bearing minerals. Commercially, osmium metal is produced as a by-product of the extraction of nickel – but less than 1 tonne of the metal is produced globally each year.

Together with its close relation ruthenium, osmium is a member of the platinum group of metals (see page 70), and the two have very similar chemical and physical properties and applications. Both metals are practically unworkable by metalworkers. The prime application for osmium is in alloys; osmium alloys are commonly used in the nibs of fountain pens, in pivots for scientific instruments and in the casings of some armour-piercing shells. Osmium alloys were once the material of choice in record player styluses. Alloyed with platinum, osmium is used to make the casing of some surgical implants, including pacemakers. Like all the platinum group elements, osmium is a widely used catalyst in the petrochemical and pharmaceutical industries.

Although the metal itself is not toxic, if osmium is exposed to the air in powdered form, it slowly reacts with oxygen to form osmium(VIII) oxide (OsO_4). This compound is toxic, and will cause damage to skin, eyes and lungs even in tiny concentrations. Osmium(VIII) oxide is volatile, and has a pungent smell; the element's name is in fact derived from the Greek word *osme*, meaning "smell". It was by studying this odorous oxide, in the residue left behind after dissolving a platinum mineral in strong acids, that English chemist Smithson Tennant discovered osmium, in 1803. (At the same time, Tennant also discovered iridium.) Despite its toxicity to the body, osmium (VIII) oxide is used to stain fatty tissues in electron microscopy, providing better contrast. Because of its affinity for fats, osmium oxide has also been used in fingerprint detection.

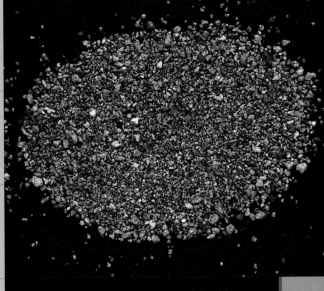

Above: Pile of small pieces of osmiridium, a natural alloy of the elements osmium and iridium.

Below: Coloured chest X-ray of a woman with a pacemaker. An osmium-platinum alloy is typically used for the electrodes of pacemakers, because of its resistance to corrosion. The electrodes supply electrical signals to the heart muscle.

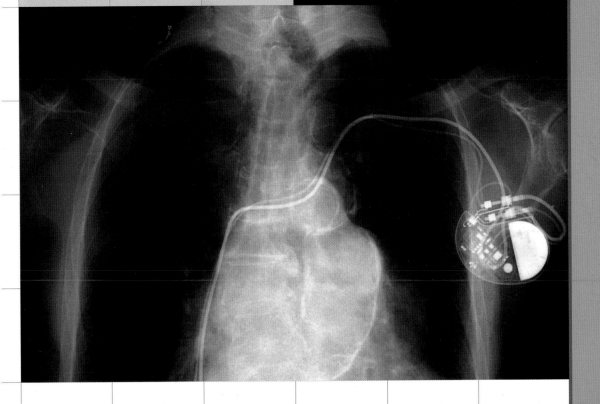

27

Co

Cobalt

ATOMIC NUMBER: 27

ATOMIC RADIUS: 135 pm

OXIDATION STATES: -1, +1, **+2**, **+3**, +4, +5

ATOMIC WEIGHT: 58.93

MELTING POINT: 1,495°C (2,723°F)

BOILING POINT: 2,900°C (5,250°F)

DENSITY: 8.85 g/cm³

ELECTRON CONFIGURATION: [Ar] 3d⁷ 4s²

Group 9 of the periodic table is populated by the transition metals cobalt, rhodium and iridium. The other element in this group, meitnerium, is unstable and radioactive and has an atomic number higher than uranium's. It is therefore in the section on transuranium elements on pages 153–7.

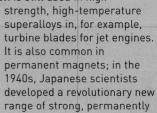

As a bulk element, cobalt is a shiny, hard metal similar in appearance and behaviour to its next-door neighbours in the periodic table: iron to the left and nickel to the right. These three are the only elements that are ferromagnetic (they can be magnetized and are attracted to a magnet) at room temperature. Ferromagnetic substances – compounds as well as elements – lose their magnetic properties above a certain temperature, called the Curie temperature. Cobalt has the highest Curie temperature of all the known ferromagnetic substances: 1,131°C.

In terms of abundance in Earth's crust, cobalt ranks about 30th. Atoms or ions of cobalt occur in several different minerals, but the pure element never occurs naturally. Cobalt is an essential element for all animals – its main role being in vitamin B12, the only one of the vitamins that contains metal atoms.

There are few concentrated deposits of cobalt, so the metal and its compounds are normally obtained as a by-product of the extraction of other metals, usually copper or nickel. Cobalt compounds were used routinely long before the element's identity was known, in brilliant blue glazes on pottery, in coloured glass and enamels, and in paints. The name cobalt comes from *Kobold* – a sprite in Germanic mythology that could be helpful or obstructive. German miners used the related term *Kobolt* to refer to an ore that was difficult to smelt, which they believed was placed in the mines by a malicious gnome. This mineral was used in making a blue pigment called zaffre. It was in this pigment that Swedish glassmaker Georg Brandt discovered and isolated cobalt, around 1735 – having shown that the blue colour was not caused by bismuth, as most people had previously believed.

Cobalt had few applications beyond pigments until early in the twentieth century. In 1903, American entrepreneur Elwood Haynes developed a range of very strong, corrosion-resistant alloys of cobalt and chromium, which he named Stellite®. Today, most cobalt is still used in high-strength, high-temperature superalloys in, for example, turbine blades for jet engines. It is also common in permanent magnets; in the 1940s, Japanese scientists developed a revolutionary new range of strong, permanently

Top: Elemental cobalt metal.

Left: Nineteenth-century morphine bottle. The blue colour of the glass is due to the presence of cobalt, in the form of cobalt oxide, which is added during the manufacture.

magnetic materials called Alnico®. These magnets are an iron-based alloy containing aluminium, nickel and cobalt (hence the name), and are still found in many applications, from electric motors to guitar pickups. In the 1970s, an even stronger range of magnetic materials was introduced – samarium-cobalt (SmCo) alloys. In cutting tools with durable and abrasive tungsten carbide blades, the tungsten carbide particles are normally embedded into cobalt metal. Cobalt compounds are used in rechargeable batteries. They are also used by the chemical industry, as catalysts in the petrochemical and pharmaceutical industries and as pigments in ceramics and plastics.

There is only one naturally occurring isotope of cobalt: cobalt-59. Another isotope, cobalt-60, is radioactive and undergoes beta decay (see pages 9–10), and as part of the process, it emits powerful and penetrating gamma-ray photons. Cobalt-60 is used as a reliable source of gamma radiation in radiotherapy (known as radiation therapy in the USA), in sterilization of medical equipment and, in countries where it is permitted, in sterilization of foods. This isotope has a half-life of just over five years, and is routinely produced in research nuclear reactors, by bombarding cobalt-59 with "slow" neutrons (see pages 153–7).

Above: Computer "stick" model of a molecule of vitamin B12, also known as cobalamin. At the centre is a cobalt ion (purple); the other colours represent carbon (yellow), hydrogen (white), nitrogen (blue) and oxygen (red).

45

Rh

Rhodium

ATOMIC NUMBER:	45
ATOMIC RADIUS:	135 pm
OXIDATION STATES:	-1, +1, +2, **+3**, +4, +5, +6
ATOMIC WEIGHT:	102.91
MELTING POINT:	1,965°C (3,570°F)
BOILING POINT:	3,697°C (6,686°F)
DENSITY:	12.42 g/cm³
ELECTRON CONFIGURATION:	[Kr] 4d⁸ 5s¹

The name "rhodium" has the same origin as "rhododendron": the Greek word *rhodon*, meaning "rose". English chemist William Hyde Wollaston named the element, after he extracted the pure metal from a rose-coloured solution he had made from a platinum ore, in 1803. Rhodium is one of the six platinum group metals (see page 70). It has properties and applications similar to the other members of that group: it is used in a variety of alloys and catalysts. The main use for rhodium since the 1970s has been in catalytic converters on vehicle exhausts, which help convert noxious carbon monoxide, hydrocarbons and oxides of nitrogen into less harmful substances.

Rhodium often occurs uncombined, in its elemental state, but it is a very rare and precious metal – it is normally the most expensive metal, several times the price of gold. Some items of silver jewellery are treated with a "rhodium flash" – a thin electroplated layer of rhodium – to improve their lustre. In 1979, the *Guinness Book of World Records* presented English singer-songwriter Paul McCartney with a rhodium-plated disc, to commemorate his status as the best-selling artist of all time.

Below: Blob of re-melted rhodium metal.

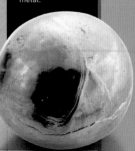

77

Ir

Iridium

ATOMIC NUMBER: 77

ATOMIC RADIUS: 135 pm

OXIDATION STATES: -3, -1, +1, +2, **+3**, **+4**, +5, +6, +7, +8

ATOMIC WEIGHT: 192.22

MELTING POINT: 2,447°C (4,435°F)

BOILING POINT: 4,430°C (8,000°F)

DENSITY: 22.55 g/cm^3

ELECTRON CONFIGURATION: [Xe] 4f^{14} 5d^7 6s^2

Iridium is one of the six platinum group metals (see page 70), and has properties and applications similar to the other members of that group. As with most of the members of the platinum group, it was discovered when chemists began subjecting platinum ores to chemical analysis early in the nineteenth century. Iridium was discovered in 1803, by English chemist Smithson Tennant, at the same time as he discovered another of the platinum group, osmium. Because the new element produced many coloured compounds, Tennant named it after Iris, the Greek goddess associated with the rainbow.

Like rhodium (see page 65), directly above it in Group 9, iridium often occurs in its elemental state, as well as in certain minerals. It is normally very rare in Earth's crust, but is more abundant in the mantle and the core, and much more abundant in the Solar System at large. Iridium arrives continuously on Earth, a component part of most meteorites. The density of pure iridium is predicted to be the highest of all the elements, but its measured density is very slightly lower than that of osmium (see page 62).

Iridium is used in a variety of alloys and catalysts – in particular where resistance to high temperature and wear is required, in the electronics, aerospace and chemical industries. Like rhodium, iridium is often used in catalytic converters on cars and other vehicles.

In the 1980s, iridium provided a vital clue in an incredible scientific detective story. Geologists had been unable to work out what caused a mass extinction that occurred about 65 million years ago – an event that spelled the end of the dinosaurs. American physicist Luis Alvarez found much higher than expected levels of iridium in rocks around the world that were laid down at that time. This iridium could only have come from out in space, where the element is much more abundant than on Earth (and the ratio of iridium isotopes present was as it is in space, but not on Earth), and Alvarez reasoned that it was due to an asteroid about 10 kilometres in diameter hitting Earth. The impact would have thrown material high into the atmosphere, blocking out sunlight and thrusting our planet into a period of desolate winter – enough to cause a mass extinction. Despite initial criticism of Alvarez's hypothesis, plenty of evidence has been collected that agrees with it – including the footprint of a colossal crater centred just off the coast of Mexico – and the impact hypothesis is now widely accepted.

Above right: Iridium metal.

Below: The K–T boundary, a thin layer of iridium-rich clay laid down 65 million years ago, as a result of the impact of a huge asteroid. Rocks from the Cretaceous (K) period are older and are below the boundary, while younger rocks from the Palaeogene period (previously called Tertiary, T) lie above the boundary.

28
Ni
Nickel

ATOMIC NUMBER:	28
ATOMIC RADIUS:	135 pm
OXIDATION STATES:	-1, +1, **+2**, +3, +4
ATOMIC WEIGHT:	58.69
MELTING POINT:	1,453°C (2,647°F)
BOILING POINT:	2,730°C (4,945°F)
DENSITY:	8.90 g/cm³
ELECTRON CONFIGURATION:	[Ar] 3d⁸ 4s²

Above right: Elemental nickel metal.

Below: Coloured scanning electron micrograph of foamed nickel metal (magnification approximately 20x). Because of the huge surface area of the nickel in this form, foams like this are used as electrodes in batteries in some electric vehicles.

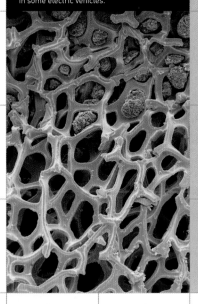

Group 10 of the periodic table includes the transition metals nickel, palladium and platinum. Also present is the element darmstadtium. Since it is unstable, radioactive and has an atomic number higher than uranium's, darmstadtium features in the section on the transuranium elements (pages 153–7).

The element nickel is named after the same mythical goblin-like sprite of Germanic mythology as the element cobalt (see page 64) – albeit by an alternative name, *Nickel*. Miners in the German town of Schneeberg found a reddish-brown ore that appeared similar to copper ore, but it was difficult to handle and, for all the miners' efforts in mining it, contained no copper. Believing it to be the work of demons, the miners called it *Kupfernickel*. When, in 1751, Swedish mineralogist Axel Cronstedt dissolved *Kupfernickel* in acid, he obtained a green solution, as you might with copper. But if you place iron in a solution containing copper, some copper will be deposited on it – and this didn't happen. Cronstedt realized that an unknown element must be present, and he managed to extract a whitish metal, which he called *Kupfernickel*, or "for the sake of convenience, nickel". Until a purer sample of the metal was extracted in 1775, by Swedish chemist Torbern Bergman, most scientists believed it was a mixture of several already-known metals.

It is fitting that copper features in the story of nickel's discovery, because there is an important range of alloys, called cupronickel, made with just these two elements. Most famously, the US coin that goes by the name "nickel" is made of cupronickel alloy consisting of 75 per cent copper and 25 per cent nickel. The US dime, quarter dollar and half dollar are also made of cupronickel, but with a smaller proportion of nickel (8.3 per cent). Several other countries have coins made of cupronickel alloy, too, and some even have coins made of nearly pure nickel. Other cupronickel alloys are found, for example, in vehicle braking and cooling systems, ships' hulls and propellers and the legs of oil rig platforms.

Adding zinc into the mix with copper and nickel produces an alloy called nickel silver – which, despite the name, contains no silver. This corrosion-resistant alloy is used to make zips and cheap jewellery. It is also commonly used to make musical instruments; "silver" saxophones and trumpets, some cymbals, and guitar frets are usually nickel silver. As a strong, stiff but flexible wire, nickel silver is popular with model makers and craft jewellery makers – and because of its corrosion resistance, fine nickel silver wire is used to anchor bristles in toothbrushes and paintbrushes.

Electroplating nickel silver with a thin layer of actual silver results in a material called electroplated nickel silver (EPNS), which looks like sterling silver. Around the end of the nineteenth century, EPNS was popular for making cutlery. It has been superseded by stainless steel in that application, but it is still popular for making decorative silver plate items. More recently, items of metal and even plastic that require protection from corrosion or just a decorative lustre can receive a layer of nickel or nickel alloy without the use of electricity, by carefully controlled chemical reactions. This process, called electroless nickel plating, is commonly used to coat door furniture, kitchen utensils and bathroom fittings.

Nickel's abundance in Earth's crust is very different from its abundance in planet Earth overall – as is the case for iron (see page 58), and for the same reason. In the very early history of our planet, molten nickel, with molten iron, dropped down to the planet's core, forcing the lighter elements up to the crust. As a result, nickel is the twenty-fourth most abundant element in the crust, but fifth overall. Nickel is also routinely delivered to Earth from space – in around one in 20 meteorites, in combination with iron. Before the Iron Age, metalworkers used meteoric iron, which had rather good anti-corrosion properties because it is actually an iron-nickel alloy.

Nickel has another similarity with iron, and with its other neighbour, cobalt: it is ferromagnetic (it can be magnetized, and is attracted to a magnet) at room temperature. Along with cobalt and the non-ferromagnetic aluminium, nickel appears in the alloy used to make cheap permanent magnets, Alnico®, which was invented in the 1940s.

Nickel is used in a wide range of other alloys beside cupronickels and Alnico® – in particular, stainless steel. The use of nickel in steel accounts for about 60 per cent of the demand for the metal. In combination with chromium, nickel is also used to make nichrome wire, which is used as the heating element in hairdryers and electric toasters. Nickel metal is the base for many high-performance

superalloys, which are typically used in jet and rocket engines. Nickel is also used as a catalyst in the oil, pharmaceutical and food industries – in particular, in the process of hydrogenating vegetable oils.

Although nickel's main applications involve the metal itself, in alloys or its pure state, some of nickel's compounds also have important applications. Most prominently, they are used in rechargeable cells (and "batteries" of cells). The nickel-cadmium (NiCad) cell was originally invented in 1899 and became very popular with the rise of consumer electronics gadgets. With a greater capacity for storing electrical energy, and an absence of toxic cadmium, the nickel metal hydride (NiMH) cell (or battery) has overtaken NiCads as the most popular rechargeable battery for portable electronics and most other applications. Both types of cell contain the nickel compound nickel(III) oxide hydroxide (NiO(OH)).

Some nickel compounds – often involving other transition metals, such as titanium – are used in a range of pigments in paints, plastics, textiles and cosmetics. Their use has recently been restricted in some countries, because of evidence that some nickel compounds may cause allergic reactions or be toxic and even carcinogenic. In terms of biology, humans and other animals have no need for nickel, but some plants and bacteria need a tiny amount.

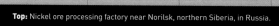

Top: Nickel ore processing factory near Norilsk, northern Siberia, in Russia.

Left: An electric toaster. The heat to brown the bread is provided by a long, thin wire made of nichrome, an alloy of nickel and chromium that is resistant to oxidation at high temperatures.

Right: Nickel metal hydride rechargeable batteries. The positive electrodes of these batteries are made of nickel hydroxide (Ni(OH)₂). The same compound is also used as the positive electrode in nickel-cadmium (NiCad) rechargeable batteries.

46

Pd

Palladium

ATOMIC NUMBER: 46

ATOMIC RADIUS: 140 pm

OXIDATION STATES: +2, +4

ATOMIC WEIGHT: 106.42

MELTING POINT: 1,552°C (2,825°F)

BOILING POINT: 2,967°C (5,372°F)

DENSITY: 12.00 g/cm³

ELECTRON CONFIGURATION: [Kr] 4d¹⁰

Like most of the other platinum group metals (see page 70), the element palladium is extremely rare, and is normally found in platinum ores. It was discovered in 1802, by English chemist William Hyde Wollaston, in a solution he created by dissolving a platinum mineral in strong acids. He named it after a new planet, called Pallas, which was discovered in the same year. (Pallas turned out to be a large asteroid, not a planet. Its name is the name of the Greek goddess of wisdom, Pallas Athena.) Wollaston also discovered another platinum group element, rhodium (see page 65), at the same time. Platinum group metals have a powerful ability to catalyse (speed up) certain chemical reactions. More than half of all palladium produced is used in catalytic converters for vehicle exhausts, and palladium catalysts are also used to produce nitric acid in the manufacture of artificial fertilizers.

Hydrogen diffuses into palladium metal better than into any other material: in 1866, Scottish chemist Thomas Graham discovered that palladium can hold more than 500 times its own volume of hydrogen. Today, a palladium-silver alloy is used as a diffusion membrane to produce ultra-pure hydrogen if required – in the semiconductor industry, for example.

Palladium is used in a range of other alloys, in particular in dentistry, where its inertness is very important. So-called "white gold", used for making rings and other jewellery, is an alloy of gold with palladium (or, alternatively, with nickel or manganese). More than 10 per cent of palladium is used in capacitors commonly fitted in mobile phones and notebook computers.

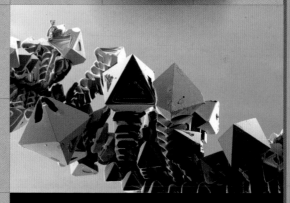

Top: Globules of palladium metal.

Above: Scanning electron micrograph of palladium metal crystals.

Left: Catalytic converter from a car's exhaust system. Inside, palladium acts as a catalyst in the breakdown of noxious substances present in the car's exhaust.

78

Pt

Platinum

ATOMIC NUMBER: 78

ATOMIC RADIUS: 135 pm

OXIDATION STATES: **+2**, **+4**, +5, +6

ATOMIC WEIGHT: 195.08

MELTING POINT: 1,769°C (3,216°F)

BOILING POINT: 3,827°C (6,920°F)

DENSITY: 21.45 g/cm³

ELECTRON CONFIGURATION: [Xe] 4f¹⁴ 5d⁹ 6s¹

Above right: Platinum (Pt), a native element, Russia.

Below: Large circular gauze made of rhodium-platinum alloy, used as a catalyst in the reaction of ammonia (NH_3) gas with oxygen (O_2) to form nitric oxide (NO).

Today, platinum is best known as a precious metal – worth considerably more than gold, and popular in jewellery for its whitish-silver lustre and its durability. In parts of what is now South America, people have been using platinum in decorative items for hundreds of years. But when Europeans first came across this inert metallic element in the mud and sand of the Rio Pinto river, they discarded it, believing it was a poor relation to silver, or perhaps "immature" gold. They even closed down gold mines that were "contaminated" with platinum. The name platinum comes from the Spanish word for silver, *plata*. Spanish conquistadors named the silvery metal *platina del Pinto* – the "little silver" of the Rio Pinto.

It was not until the middle of the eighteenth century that scientists began to take note of platinum. The English ironmaster and assayist Charles Wood collected a sample of platinum in Cartagena, New Spain (now Colombia), and sent it to England for analysis in 1741. Also in the 1740s, Spanish mathematician and explorer Antonio de Ulloa recorded details of the metal that he had examined in Peru. In 1752, Swedish assayist Henrik Scheffer studied platinum, and identified it as an element and a precious metal.

Scientists became very interested in platinum over the next 50 years, submitting it to all kinds of physical and chemical analysis. Platinum is the least reactive of all metals, so it did not give up its secrets easily. Its inertness was put to good use in 1799, when the French Academy of Sciences made a platinum bar to act as the standard metre length and a cylinder of platinum to act as the standard kilogram. Today, the metre is defined by more sophisticated means, but an updated platinum cylinder (alloyed with 10 per cent iridium), made during the 1880s and officially sanctioned on 28 September 1889, is still the international reference for the kilogram today. It is held at the Bureau International des Poids et Mesures, in Paris, France.

When English chemist William Hyde Wollaston and his friend, English chemist Smithson Tennant, went into business together in 1800, one of their aims was to find a way to produce very pure platinum. In 1803, Wollaston and Tennant dissolved platinum in a mixture of concentrated

nitric and hydrochloric acid, called aqua regia. Wollaston studied the resulting solution, which contained dissolved platinum and two other previously unknown elements: palladium and rhodium. Tennant studied the black residue left behind after the reaction, and discovered two more elements, iridium and osmium. Together with platinum and another element, ruthenium, discovered 40 years later, these elements make up what is known as the platinum group of metals. The six elements of the platinum group are all very rare and normally occur together. They have very similar properties – in particular, their high malleability and very low reactivity – and applications.

As is the case for all the platinum group elements, platinum is one of the least abundant elements in Earth's crust. It is normally obtained as a by-product of the extraction of nickel or copper, mostly in South Africa, Russia and Canada. Platinum is used as a catalyst to speed up several industrial chemical reactions, including the manufacture of nitric acid from ammonia. The use of platinum in catalytic converters in vehicle exhaust systems (see page 65) accounts for more than half the demand for the metal. Platinum metal is also found on the surface of computer hard disks, where it improves the density of data that can be stored there. Platinum is as malleable and as ductile as gold, and it can be made into extremely fine wires and extremely thin sheets. Missile nose cones are sometimes coated with a fine sheet of platinum. Its inertness makes platinum ideal for a coating on vehicle spark plugs, and in medical applications – although the high price of the metal restricts the use to tiny electrodes in, for example, pacemakers. Some dental bridges are made of platinum alloyed with gold, copper or zinc.

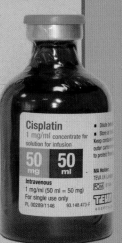

Certain compounds of platinum can bind on to certain sites along the length of DNA molecules, making it impossible for the DNA to replicate. This property has been put to use in medicines that can inhibit the growth of tumours. The first, and best known, of these platinum anti-cancer drugs is cisplatin, or *cis*-diamminedichloroplatinum(II), which has been used against a range of cancers since the late 1970s.

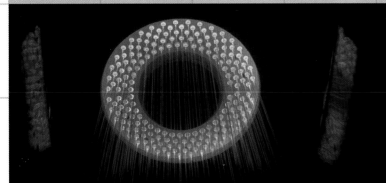

Top: The Atok platinum mine in the Limpopo Province of South Africa. South Africa has the second largest platinum reserves, after Russia.

Above: Treatment-ready concentrate of cisplatin chemotherapy drug. Each molecule of the active component contains an atom of platinum.

Left: Thin glass fibres emerging through tiny holes in a bushing made from a platinum-rhodium alloy. This alloy is chosen because molten glass "wets" it, as water wets glass, and it is very stable at high temperatures.

29

Cu

Copper

ATOMIC NUMBER: 29	
ATOMIC RADIUS: 135 pm	
OXIDATION STATES: +1, **+2**, +3, +4	
ATOMIC WEIGHT: 63.55	
MELTING POINT: 1,085°C (1,985°F)	
BOILING POINT: 2,656°C (4,650°F)	
DENSITY: 8.94 g/cm³	
ELECTRON CONFIGURATION: [Ar] 3d¹⁰ 4s¹	

Right: A nugget of fairly pure "native" copper.

Below: Copper is malleable – it can be beaten or squeezed into sheets. These rolls also exhibit the orange colour of this element – unusual among metals.

Group 11 of the periodic table includes three transition metal elements that often occur naturally in their "native" state, uncombined with other elements: copper, silver and gold. It also includes the radioactive element roentgenium, which is not found in nature, and which has an atomic number higher than uranium's. Roentgenium is therefore featured in the section on transuranium elements, on pages 153–7.

Copper is the twenty-sixth most abundant element in Earth's crust. Because it occurs as a native element, copper was known to ancient people; the oldest known artefacts made of copper are about 10,000 years old. Copper was first smelted from its ores around 7,000 years ago, and first combined with tin – making bronze – around 5,000 years ago. Unlike copper, bronze is hard and can be given a durable sharp edge. Across much of Europe and Asia, the Bronze Age began around 3000 BCE and ended around 1000 BCE, when the alloy was largely superseded by iron and steel. Bronze is still used today, to make sculptures and medals; it is also used to make bells, for high-quality cymbals, and as the winding around guitar and piano strings.

Another ancient alloy of copper with a wide range of uses is brass: copper with zinc (see page 78). Both brass and bronze were popular in the Roman Empire. The name copper is derived from *Cuprum*, Latin name for Cyprus, where the Romans mined most of their copper. Copper is one of the seven "metals of antiquity" (see page 76); in Ancient Greek and Roman mythology, and in alchemy, copper was associated with Venus.

There is a wide range of other copper alloys in use today, in addition to bronze and brass. Coins in many countries are made from cupronickel alloys (see page 67). In situations where resistance to corrosion is important, cupronickel or copper-aluminium alloys are used – for example, in cars, ships, aircraft landing gear and oil rigs. Copper is tightly bonded to steel – but not alloyed with it – in a device called a bimetallic strip, which bends in response to temperature changes. Bimetallic strips are used in some thermometers, thermostats and as safety devices that can automatically break a switch if too much current flows through a circuit. They work because copper expands more than steel does as the temperature increases, making the strip bend.

Copper is one of the only two metals whose lustre is not silver-grey, the other being gold. Both copper and gold absorb photons of light at the higher-energy blue end of the spectrum, while all other metals reflect all photons of light more or less equally. Copper normally has a bright orange lustre, but if it is left exposed to the air, it reacts slowly with oxygen, forming a dull coating of copper(II) oxide (CuO). Left even longer, a copper surface will take on a fragile greenish coating called verdigris, mostly copper(II) carbonate ($CuCO_3$), as a result of a reaction between the copper, oxygen and carbon dioxide. The green colour of the Statue of Liberty – the largest copper statue in the world – is the result of verdigris.

The main use of copper is as electrical wiring, including the windings of electromagnets, transformers and motors. Only silver has a better electrical conductivity at room temperature, but it is too expensive to be used for domestic wiring. Copper is both unreactive and easy to work with – it can readily be cut with a steel blade, and beaten, bent or soldered – which is why it is a good choice for use in pipework for water and gas supplies and heating systems. Altogether, a typical modern home contains around 200 kilograms of copper and a typical car about 20 kilograms.

The biological importance of copper in animals first came to light in 1928, when researchers at the US Department of Agriculture found that rats need copper as well as iron to make haemoglobin (see page 60 [iron]), the pigmented substance that carries oxygen in the blood. In rats – and humans – copper is present in enzymes involved in building the (iron-containing) haemoglobin molecule. Research since then has led to the discovery of copper's role in several more enzymes and other proteins. (Octopuses and many other molluscs, for example, use a blue-coloured, copper-based protein called haemocyanin, rather than a red, iron-based one, for transporting oxygen; they have blue blood as a result.) Copper is an essential element in humans; an adult requires around 1 milligram per day. Good dietary sources of copper include liver, egg yolk, cashew nuts and avocado. An adult human body contains about one-tenth of a gram of copper – around 90 per cent of it in a blood protein called ceruloplasmin.

Copper-rich alloys are being used increasingly on surfaces in hospitals and schools, because copper has a proven antimicrobial action. Many pathogenic bacteria, viruses and fungi can survive for days or even weeks on various surfaces, but they are killed or deactivated within a matter of hours on copper surfaces. In tests, infection rates in hospitals have been greatly reduced. Copper alloys are also commonly used to make netting in fish farms; the copper prevents the accumulation of plants, algae and microorganisms, and improves the health of the fish as a result.

Above: A fifteenth-century astrolabe made of brass, a copper-zinc alloy. Brass has a much lower melting point than either of its constituents, but is easy to work with and is resistant to corrosion.

Below: Reels of copper wire. Copper has the second highest electrical conductivity of all metals, after silver, and copper wire is the most common choice for electrical wiring.

47

Ag

Silver

ATOMIC NUMBER: 47

ATOMIC RADIUS: 160 pm

OXIDATION STATES: **+1**, +2, +3

ATOMIC WEIGHT: 107.87

MELTING POINT: 962°C (1,764°F)

BOILING POINT: 2,162°C (3,924°F)

DENSITY: 10.50 g/cm³

ELECTRON CONFIGURATION: [Kr] 4d¹⁰ 5s¹

Unlike its Group 11 partners copper and gold – but like all other metals – the element silver has a silvery lustre. Within a sample of the pure metal, the electrons associated with silver atoms are free, and they can effectively reflect pretty much any photon that comes their way. This freedom has another important effect: both heat and electricity are transferred through solid matter by the movement of electrons, and the free electrons in silver give it a greater conductivity than any other material at room temperature.

Above: A miner installing supports in a silver mine. Although silver is sometimes found native, as nuggets, it is more commonly extracted from ores also containing copper, nickel and lead.

Above right: Pure silver.

Silver is found in native (uncombined) form, so it was known to ancient metalworkers. It is one of the seven "metals of antiquity" (see page 76); the chemical symbol for silver, Ag, derives from its Latin name, *argentum*. In Ancient Greek and Roman mythology, and in alchemy, silver was associated with the Moon. In 2010, NASA deliberately crashed a space probe into a crater on the dark side of the Moon, to produce a plume of debris they could study from Earth. Along with other elements, the scientists were surprised to find silver as a tiny but significant component of the plume. The apparent brightness of the Moon has nothing to do with any reflective silver particles in the Moon's surface, however. In fact, the Moon's soil is dark grey, and reflects only 12 per cent of the light that falls on it; it simply looks bright against the dark night sky.

Like all the metals found native, silver is generally unreactive. Silver is resistant to attack by water or oxygen, for example. But silver does react with hydrogen sulfide (H_2S) in the air, forming a tarnish of silver(I) sulfide (Ag_2S) that can be removed by polishing. Most ores that contain silver are sulfide minerals – but silver is also found combined with arsenic and antimony. Around 20,000 tonnes of silver is produced worldwide each year, most of it from silver mines, and some as a by-product of the refining of mined copper, gold and lead. The biggest producers are Mexico and Peru. A further 7,000 tonnes or so is produced by recycling silver from scrap.

In addition to the long-standing use of silver in jewellery, decorative items and coins, one of the major uses for silver in the twentieth century was photography. When exposed to light, certain silver compounds blacken, as silver metal precipitates out. A German anatomist called Johann Heinrich Schulze, in 1727, was the first person to prove that this behaviour was due to light – and not heat or exposure to air, as others had believed. In 1801, German physicist Johann Ritter used this property of silver(I) chloride (AgCl) to discover ultraviolet radiation. Inspired by the discovery of invisible infrared radiation beyond the red end of the spectrum a year earlier, Ritter placed paper impregnated with silver chloride at the violet end of the spectrum created by sunlight passing through a prism. He found that blackening still occurred just beyond the violet end of the spectrum.

In photography, silver(I) nitrate ($AgNO_3$) is used as a precursor to silver halide compounds, such as silver(I) iodide (AgI – see iodine, page 143), which are used on photographic films. The first person to produce photographic images was French inventor Nicéphore Niépce, who produced a few images in the 1820s. Niépce initially used bitumen, which hardens slowly in the light; looking for something quicker, he and his compatriot Louis Daguerre experimented with silver compounds. Photography gradually developed through the nineteenth century, culminating in cheap, portable "point-and-shoot" cameras made possible by mixing the silver compounds in a gel pasted on to celluloid. This innovation also led to the development of the film industry – and, fittingly perhaps, silver metal was embedded in the projection screens in early cinemas, a fact that led to cinema being referred to as the "silver screen".

With the rise of digital cameras, the demand for silver by the film and photographic industry has plummeted in the twenty-first century. But another application of silver is growing rapidly: solar cells. Silver is often used, suspended in a paste, to make fine connecting wires. For example, the characteristic metal grid on the front of solar cells, which allows the electric current to flow, is normally silver, applied in this paste form. In the electronics industry, silver is used in switches – especially those found on telephone keypads and computer keyboards – and is applied as an ink or film to form electrical pathways on printed circuit boards. Silver foil is also used to make the antennas in radio frequency identification tags (RFID), which are commonly attached to clothes, books and other consumer goods to prevent theft.

In the nineteenth century, electroplating became a popular way to produce affordable silver items,

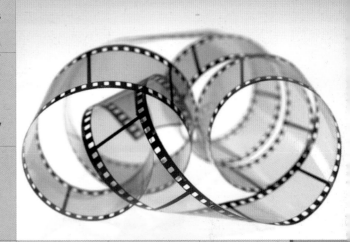

including "silver plate", and "electroplated nickel silver" (EPNS) cutlery (see page 67). In the 1890s, scientists even used electroplating to define the unit of electric current, the ampere, in terms of the current that would deposit 0.001118000 grams of silver per second from a solution of silver nitrate. (Today, the definition of the ampere is based on the number of electric charges flowing per second.)

Silver has long had applications in medicine; even the Greek physician Hippocrates noted its healing and preventative properties. Like copper, silver is quite toxic to micro-organisms, but not to humans. Silver compounds are used in some bandages, and silver nanoparticles are used in creams to treat skin infections. And, in the coming decades, silver may also play a major role in treating cancer. In 2012, a team of scientists at the University of Leeds, UK, announced that certain silver compounds have a similar potency against cancer cells to one of the leading cancer drugs, cisplatin (see page 71) – but without the side effects of cisplatin, which are due to platinum's toxicity to humans. Silver is also used in some water purification filters, as dissolved silver ions that can kill bacteria and prevent the build-up of algae.

Above: A strip of photographic film. The surface of the film is coated with an emulsion containing light-sensitive silver compounds. When light falls on them, they produce elemental silver, which darkens the emulsion.

Left: Macrophotograph of a wound dressing impregnated with small particles of silver, which are toxic to pathogenic (disease-causing) microbes.

79

Au

Gold

ATOMIC NUMBER: 79

ATOMIC RADIUS: 135 pm

OXIDATION STATES: -1, +1, +2, **+3**, +5

ATOMIC WEIGHT: 196.97

MELTING POINT: 1,064°C (1,948°F)

BOILING POINT: 2,840°C (5,144°F)

DENSITY: 19.30 g/cm³

ELECTRON CONFIGURATION: [Xe] 4f¹⁴ 5d¹⁰ 6s¹

Like its Group 11 partners copper and silver, the element gold is often found native – in other words uncombined with other elements. This is because gold is very unreactive: it is not attacked by water or oxygen, for example, even when hot. It does dissolve in a mixture of nitric and hydrochloric acid, called aqua regia – but unlike all other metals except platinum, it does not dissolve in nitric acid alone. A reliable and long-established test for pure gold involves applying a drop of nitric acid; this was the origin of the term "acid test", which first became popular among gold prospectors and assayers in the 1890s.

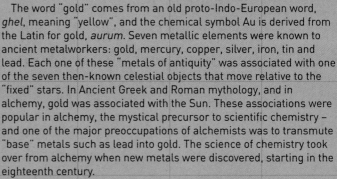

Because it exists in its native state, gold was known far back into prehistory; the oldest known gold artefacts, discovered in a burial site near Lake Varna, Bulgaria, are 6,000 years old. In many early civilizations, gold had little practical value – it is too soft to make tools and weapons. Instead, its yellow lustre and rarity led to its being adopted for decoration and as a symbol of wealth.

The word "gold" comes from an old proto-Indo-European word, *ghel*, meaning "yellow", and the chemical symbol Au is derived from the Latin for gold, *aurum*. Seven metallic elements were known to ancient metalworkers: gold, mercury, copper, silver, iron, tin and lead. Each one of these "metals of antiquity" was associated with one of the seven then-known celestial objects that move relative to the "fixed" stars. In Ancient Greek and Roman mythology, and in alchemy, gold was associated with the Sun. These associations were popular in alchemy, the mystical precursor to scientific chemistry – and one of the major preoccupations of alchemists was to transmute "base" metals such as lead into gold. The science of chemistry took over from alchemy when new metals were discovered, starting in the eighteenth century.

Most gold exists in nature as tiny particles embedded in rocks, often mixed with silver. Where rivers erode these rocks, particles of gold wash downstream; as they do so, they can accumulate into flakes and nuggets. This is why people pan for gold – collecting gravel and silt from a stream in a round-bottomed pan and agitating it, so that any gold present falls to the bottom and less dense materials can be tipped out. This is really a hobby, although occasionally fairly large nuggets can be found. The majority of gold is extracted by mining – the biggest producers being China, Australia and the USA. According to the World Gold Council, an estimated 166,600 tonnes of gold had been mined from the beginning of civilization until 2012.

Above: A prospector panning for gold in river sediment. Swirling water and sediment around the pan separates particles by their density, so that the denser gold particles settle on the bottom of the pan.

Above right: Gold nugget, found by panning in a river. Nuggets like these are formed by cold welding – the accumulation of tiny flecks of gold released from the rocks of the riverbed by erosion.

In 1872, Swedish-English chemist Edward Sonstadt established that there is gold, in solution, in seawater. The concentration is extremely low – perhaps 1 milligram per cubic kilometre – but it means there are tens of thousands of tonnes in total in the world's oceans. German chemist Fritz Haber spent considerable time and resources trying to extract gold from seawater, to help pay off German debt after the First World War, but with no success. Despite many ingenious schemes, no one else has managed to obtain significant amounts of gold from the oceans since then.

In most countries, the purity of gold is measured in carats (symbol ct; in the USA it is kt or k, for karat). In theory, 24-carat gold is 100 per cent pure – although that is all but impossible in practice and a tolerance of 0.01 per cent is allowed. An alternative system is millesimal fineness, a scale on which 1,000 represents pure gold. So an 18-carat gold sample contains 75 per cent gold and is 750 fine. The purest known sample of gold ever produced – at the Perth Mint, Australia, in 1957 – had a confirmed fineness of 999.999; only one in every million atoms in the sample was not gold.

As well as being used for jewellery and other decorative items, and as gold reserve currency, gold has many practical uses. Because it is so ductile and completely resistant to corrosion, it is normally the metal of choice for fine bonding wires that connect to semiconductor chips – its biggest industrial application. High-end audio connectors are gold or gold-plated, because gold does not tarnish and has very high electrical conductivity.

Gold has long been used in dental replacements, often as an amalgam with mercury, although it is being superseded by new porcelain and synthetic composite materials. Gold is also used in electron microscopy – either as an extremely fine coating that conducts the electrons away from the sample or as tiny gold "nanoparticles" that attach to specific proteins or tiny structures within cells, giving them high contrast in electron micrograph images. Gold nanoparticles are also being used increasingly in medicine, where they can be made to enter cancerous cells and then be a target for irradiation with laser light – or can carry anti-cancer drugs directly into the cancerous cells' nuclei.

Gold is non-toxic, and is even used in small amounts in certain foods, normally beaten into extremely fine gold leaf – gold is the most malleable of all metals – or as tiny flecks. Edible gold features in the world's most expensive desserts. In 2010, for example, a New York restarant added to its menu a dessert called the Frrozen Haute Chocolate, which is topped with 5 grams of 24 carat edible gold and costs $25,000 (£15,000) – although the price includes a jewel-encrusted spoon and a diamond bracelet.

Top: The death mask of the Egyptian pharaoh Tutankhamun (1341–1323 BCE) is made of gold inlaid with precious stones and coloured glass.

Above: Coloured scanning electron micrograph showing the fine gold wires and gold contact pads connecting a computer chip to the rest of the computer circuit board.

Left: Gold ingots, each with a mass of 1 kilogram. Each ingot carries a stamp detailing the mass, purity and reference number.

30

Zn

Zinc

ATOMIC NUMBER: 30
ATOMIC RADIUS: 135 pm
OXIDATION STATES: +1, **+2**
ATOMIC WEIGHT: 65.39
MELTING POINT: 420°C (787°F)
BOILING POINT: 907°C (1,665°F)
DENSITY: 7.14 g/cm³
ELECTRON CONFIGURATION: [Ar] $3d^{10} 4s^2$

Above right: Sample of pure zinc.

Below: Crystals of sphalerite, the principal ore of zinc.

Group 12 of the periodic table is the last of the transition metal groups, and the rightmost group of the d-block (see page 38). It contains three metals – zinc, cadmium and the only metal that is liquid at room temperature, mercury. It also includes the radioactive element copernicium, which is not found in nature, and which has an atomic number higher than that of uranium. Copernicium is therefore featured in the section on the transuranium elements, on pages 153–7.

The English word "zinc" comes from the German name for the metal, *Zink*, which in turn probably comes from the German *zinke*, meaning "prong" or "point". The Swiss physician Paracelsus used the name *zinken* in the 1520s, to describe the tooth-like projections of zinc crystals. Paracelsus noted that zinc was a metal; it became the eighth metal known, after the seven metals of antiquity (see page 76).

Pure zinc is silvery-grey, but on exposure to the air it reacts with oxygen and carbon dioxide, quickly forming a thin layer of dull grey zinc(II) carbonate ($ZnCO_3$), which renders it fairly unreactive. Adding zinc granules into even a weak acid will cause bubbles of hydrogen gas to form, and the zinc to dissolve slowly in the acid. Zinc is the twenty-fourth most abundant element in Earth's crust, and occurs in a variety of minerals, including sphalerite, the most common.

Zinc is an essential element for all living things, and is found in a wide range of foods, including meat, shellfish, dairy products and cereal products. An adult requires around 5 milligrams of zinc per day in his or her diet. An estimated two billion people around the world suffer from zinc deficiency, whose main symptoms are diarrhoea, skin lesions, hair loss and loss of appetite.

Around 12 million tonnes of zinc metal is produced each year, and zinc is the fourth most used metal, after iron, aluminium and copper. More than half of it is used as a coating to protect iron and steel surfaces from corrosion, a process called galvanization. Car bodies, corrugated-iron roofing panels and barbed wire are galvanized, either by dipping them in molten zinc or by electroplating them.

Another major use of zinc is the production of small, intricate components by die casting. By forcing molten zinc into moulds, complicated shapes can be made without machining; zinc is ideal for die casting, because of its corrosion resistance and its relatively low melting point of just under 420°C. Sometimes, die-cast zinc is actually an alloy, with small amounts of other metals, such as lead, tin and aluminium. Zinc is also commonly used in alloys made into certain coins, such as the UK pound coin. The US penny coin is 97.5 per cent zinc, with a copper coating.

The best-known zinc alloy is brass, today associated with musical instruments such as trumpets. Brass is a mixture of copper and zinc, which has been used for around 3,000 years – and, in making brass, people were using zinc hundreds of years before they identified the element itself. Originally, brass was made by accident, during the smelting of copper ores that happened to contain zinc. Within a few hundred years, Roman metalworkers were producing it by heating copper with a zinc ore called calamine, which is mostly zinc carbonate ($ZnCO_3$); at high temperatures, zinc vapour was produced, which infiltrated the copper. This method was used to make the huge quantities of brass the Ancient Romans used in weapons and coins; however, Roman metallurgists were unaware that zinc is a metal in its own right. From the thirteenth century onwards, Indian metallurgists did manage to extract zinc metal in Rajasthan, by producing zinc vapour and distilling it, and brass was made there by mixing the two metals. Today, the Rampura Agucha mine in Rajasthan is the world's biggest zinc mine, from which more than 600,000 tonnes of zinc metal is produced per year.

News of the new metal spread from India to Europe, but zinc metal was not produced in large quantities there until 1738, when English chemist William Champion invented a distillation process, similar to that used in India hundreds of years earlier. A few years later, German chemist Andreas Marggraf isolated fairly pure zinc in the laboratory and studied its properties. In 1799, Italian scientist Luigi Galvani invented his "voltaic pile", a battery made of alternating discs of copper and zinc separated by moist discs of cardboard. Today, zinc metal is used as the container and the negative terminal of zinc-carbon batteries (now being largely superseded by alkaline batteries).

The most important zinc compound is zinc(II) oxide (ZnO), with around a million tonnes produced worldwide each year. This white powder was once popular among artists, as a white pigment called Chinese white. Today, it is added to plastics to protect them against damage from ultraviolet radiation; it is also used in sunscreens. Another common medical application is in

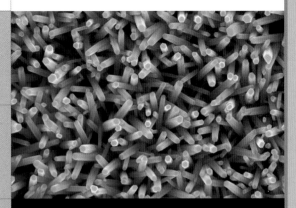

Above: Coloured scanning electron micrograph of artificially grown crystals of zinc oxide only a few nanometres (billionths of a metre) long. Zinc oxide is a semiconductor, and researchers hope that "nanowires" like this will one day be used to make a range of electronic devices.

Below: Two European oysters (Ostrea edulis). Oysters provide more zinc per standard serving than any other food.

calamine lotion, which is the pink mixture commonly used to soothe itching rashes. Calamine lotion contains a mixture of zinc(II) oxide (ZnO) and iron(III) oxide (Fe_2O_3). (The mineral calamine, which was used to make brass (see above), is actually a mixture of two zinc minerals, smithsonite and hemimorphite. It is now customary to use these two mineral names to avoid confusion with calamine lotion.) Zinc oxide is also used in the production of vulcanized rubber for tyres, and is added to lubricating oils to reduce oxidation of the metal parts of engines. In addition, it is added to fertilizers and animal feed, and is present in some dietary supplements.

Another compound, zinc(II) sulfide (ZnS), is one of the main compounds used to make "glow-in-the-dark" (phosphorescent) pigments and electroluminescent displays for consumer electronics. With tiny amounts of other compounds added, it can glow green (with copper), blue (with silver) or red (with manganese) after being exposed to light or an electric field or current.

48

Cd

Cadmium

ATOMIC NUMBER: 48

ATOMIC RADIUS: 155 pm

OXIDATION STATES: +1, **+2**

ATOMIC WEIGHT: 112.41

MELTING POINT: 321°C (610°F)

BOILING POINT: 767°C (1,413°F)

DENSITY: 8.65 g/cm³ ·

ELECTRON CONFIGURATION: [Kr] 4d¹⁰ 5s²

Element 48, cadmium, is very similar to zinc in its properties and applications, but is much less abundant. Cadmium often occurs in small amounts in zinc ores, such as calamine (see page 79), and the 20,000 or so tonnes of cadmium produced each year are produced during the extraction of zinc. In 1817, German chemist Friedrich Stromeyer was investigating a sample of calamine (zinc carbonate) that turned yellow, then orange on heating. After removing the zinc content, Stromeyer was left with a silver-blue metal that he realized was a previously unknown element. He based its name on the Latin or Greek word for the mineral calamine: *cadmia* or *kadmeia*.

Like most transition metals, cadmium produces colourful compounds. Stromeyer combined cadmium with sulfur, producing a bright yellow compound, cadmium(II) sulfide (CdS). Similarly, cadmium(II) selenide (CdSe) is a bright red colour. Varying the amounts of sulfur and selenium, it is possible to produce a range of bright colours that do not fade – and in the 1840s, artists began using cadmium pigments to great effect. These light-fast pigments are visible today in many of the great works of art by painters such as Vincent Van Gogh and Claude Monet; however, although some artists still use cadmium-based pigments, concerns over cadmium toxicity have encouraged the development of synthetic alternatives. Exposure to cadmium can provoke an inflammatory response, leading to symptoms like those of the flu: fever, respiratory problems and aching muscles, sometimes called "the cadmium blues". Higher or longer-term exposure can result in severe respiratory problems and kidney failure.

The production of cadmium was not significant until the 1930s, when demand for zinc grew in response to the need for galvanizing car bodies and aeroplane fuselages. Like zinc, cadmium is sometimes used for galvanizing, mainly as

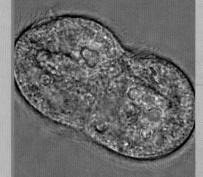

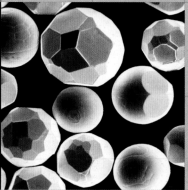

Top right: Pure cadmium, a soft, silver-blue metal.

Above: Glowing quantum dots inside a single-celled organism. A quantum dot is a tiny crystal of semiconductor (in this case, cadmium selenide) that emits precisely "tuned" photons of light. Here, they are being used to track the movement of nanoparticles through the food chain.

Right: Coloured scanning electron micrograph of crystals of fine cadmium powder, for use in solar cells or powder coating metal surfaces. The largest crystals in the picture are about 0.1 millimetres in diameter.

an electroplated layer, and mainly in the aircraft industry. In the 1970s, cadmium sulfide was used in plastics, as zinc sulfide is today. But due to concerns over cadmium's toxicity, this application has steadily waned; and in the European Union, the use of cadmium in many plastics (and in jewellery and cosmetics) was banned in the 1990s, with an outright ban coming into force in 2011. Since the 1980s, demand for cadmium has been dominated by nickel-cadmium (NiCad) rechargeable batteries, in which a cadmium plate forms the negative electrode. Other designs – notably nickel metal hydride (NiMH) batteries – have become more popular in recent years, because they are cheaper and do not contain toxic cadmium.

Right: Rechargeable nickel-cadmium (NiCad) batteries in a charger.

80
Hg
Mercury

ATOMIC NUMBER: 80
ATOMIC RADIUS: 150 pm
OXIDATION STATES: +1, **+2**, +4
ATOMIC WEIGHT: 200.59
MELTING POINT: -39°C (-38°F)
BOILING POINT: 357°C (675°F)
DENSITY: 13.55 g/cm^3
ELECTRON CONFIGURATION: [Xe] 4f^{14} 5d^{10} 6s^2

Above right: Pure drops of mercury resting on a hard surface. Mercury is the only metal that is liquid at normal room temperature and pressure.

Mercury is one of the seven "metals of antiquity" (see page 76), and in Ancient Greek and Roman mythology it was associated with the planet Mercury. The chemical symbol for mercury, Hg, is derived from the Latin word *hydrargyrum*, which means "watery silver".

Of all the metallic elements, only mercury is liquid at normal room temperature – and there is only one other element, non-metallic bromine, that shares this property. Mercury has the highest atomic number and the highest atomic mass of all the transition metals, and, like nearly all transition metals, it has a silver lustre. It is also a potent toxin (see below).

    Although mercury does occur native – uncombined with other elements – this is very rare in nature. However, it is easy to obtain mercury from naturally occurring mercury compounds. The most important ore of mercury is the orange-red mineral cinnabar, which consists mainly of mercury(II) sulfide (HgS). Roasting cinnabar dissociates this compound, producing mercury vapour that condenses as droplets of shiny metal. Cinnabar was highly prized in many ancient civilizations, as a pigment to make the luscious colour vermilion, often reserved for kings and emperors. However, people have long been aware of mercury's toxicity: cinnabar miners and roasters were prone to peculiar symptoms and very short life expectancies. In the Roman Empire, criminals

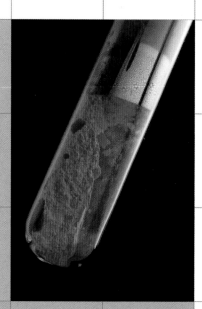

and slaves were sent to cinnabar mines in Spain and Slovenia.

Cinnabar was also highly prized by alchemists – the mystical forerunners of modern chemists. In many alchemical traditions, sulfur, mercury and salt (as ash or earth) were the "principles" of life, spirit and body in all things. When they roasted cinnabar, alchemists could separate and reunite these three principles at will – the salt was the ash left behind. Many alchemists also believed that mercury contained the metal "principle", and could be transmuted into any metal, including gold.

As alchemy began to give way to science, mercury played a vital role in many experiments and discoveries. In 1643, Italian physicist Evangelista Torricelli invented the mercury barometer; atmospheric pressure and blood pressure are still often given in units of mmHg (millimetres of mercury). The mercury-in-glass thermometer was invented by the Polish-born German scientist Daniel Gabriel Fahrenheit in 1714. In early research into electric circuits and electromagnetism, the electrical contacts often comprised wires in small cups of mercury – and when English scientist Michael Faraday made the first, crude, electric motor in 1821, the moving part was a current-carrying wire that made a stirring motion in a bowl of mercury, which kept the moving wire connected to the electricity supply. The compound mercury(II) oxide (HgO) was also crucial in the discovery of oxygen (see page 123). Also in the nineteenth century, the use of mercury in vacuum pumps allowed scientists to produce extremely low pressures, leading to the discovery of the electron and the invention of television and, in the vacuum tube, the beginning of electronics in the early twentieth century.

Mercury has the lowest conductivity – of both electricity and heat – of all the metallic elements (although there are metal alloys with lower conductivity). The atoms of most metals easily lose their outer electrons, attaining a stable electron configuration (see pages 11–12). These outer electrons are shared among all the atoms, forming a metallic bond that holds the atoms in place. The shared electrons roam freely among the atoms, and are responsible for metals' good conductivity of electricity and heat. Mercury has filled s-, d- and f-orbitals ($6s^2\ 5d^{10}\ 4f^{14}$); in this stable configuration, mercury atoms do not easily share their electrons (see pages 8–9). This also makes the bonds between the atoms much weaker, which helps explain why mercury has such a low

Above: Mercury-in-glass thermometer being used to measure the temperature of water. Once commonplace, mercury is nowadays rarely used in thermometers, due to its toxicity.

Above right: Heating mercury(II) oxide in a test tube. The compound dissociates, liberating oxygen, which escapes to the air, and mercury vapour, which condenses on the side of the tube in liquid mercury beads.

melting point. The other Group 12 metals zinc ($4s^2\,3d^{10}$) and cadmium ($5s^2\,4d^{10}$) also have relatively low conductivities and low melting points; the effect is greatest for mercury, because it has more protons in the nucleus, creating a stronger attractive force between the nucleus and electrons (see page 8).

Despite its low conductivity in liquid form, mercury vapour conducts electricity well. This is why mercury is used inside the tubes of energy-saving fluorescent lamps. As electric current flows through the vapour, electrons in the mercury atoms are excited, then release their energy as photons of ultraviolet radiation. The ultraviolet strikes phosphor chemicals in pigments on the coating of the tube that produce red, green and blue light – making an overall whitish light.

While the use of energy-saving lamps is increasing, many of the other applications of mercury in industry and in everyday life are disappearing, as a result of concerns over mercury's toxicity. For example, there is considerable controversy over the use of mercury, in a solid amalgam with gold or other metals, in dental fillings. There is disagreement over the amounts of mercury ingested from fillings, and the amount of harm they cause, but they are now banned in several countries. One particular concern is the mercury vapour that is inevitably released into the environment when bodies of those with amalgam fillings are cremated. Alternatives to amalgam are available, and have recently begun to rival amalgam's ease and low cost.

Breathing in mercury vapour and ingesting mercury compounds are the main ways this element enters the human body. Inside the body, mercury has a range of damaging effects; symptoms of chronic (long-term) mercury

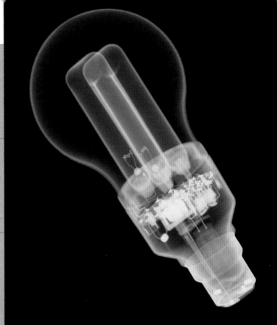

Above: X-ray image of a compact fluorescent lamp. The glass tubes inside the bulb are filled with mercury vapour, which produces ultraviolet radiation when electric current flows through it; the ultraviolet excites phosphors on the inner surface of the tubes, which emit visible light.

Below left: An amalgam is any alloy of mercury. In dental amalgam, mercury is normally alloyed with silver, tin and copper. Mercury-free resins are gradually replacing amalgam in dentistry.

poisoning include swelling, skin peeling, loss of hair, kidney malfunction, nervousness, trembling, insomnia and dementia. Mercury can accumulate in certain organs, so that its harmful effects can increase over time – and it can pass to unborn children, where it can cause congenital disorders including malformation.

The most notorious episode in the history of mercury was the poisoning of the Bay of Minamata, in Japan, in the 1950s. A nearby chemical factory produced an ion called methylmercury (CH_3Hg^+) as a waste product and released it into the water. Fish and other organisms absorbed the methylmercury, which is extremely neurotoxic (it damages the nervous system). When the local population ate the fish, hundreds of them began to suffer with strange neurological symptoms – and thousands have suffered since, many of them dying as a result. Even after the cause of the disease had been identified, it took several years before the factory stopped releasing methylmercury into the water and even longer before proper compensation was paid out to the fishermen, victims and their families.

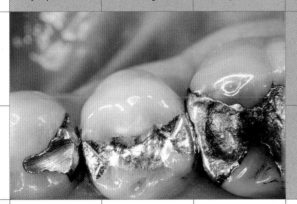

Interlude: The f-block – the lanthanoids and the actinoids

In Group 3 of the periodic table, there appear to be two gaps – empty slots below element 39, yttrium. But these slots are are far from empty: each contains fifteen elements! The "missing" elements are in a separate block – the f-block, which is normally found below the rest of the table but is, in a sense, part of Group 3. The next section of this book (pages 86 to 97) is devoted to the elements of the f-block. There are two rows in the f-block, with elements from Periods 6 and 7 of the periodic table.

The elements in the top row, from Period 6, are the lanthanoids (pages 86 to 93), sometimes called lanthanides – named after the first of them, lanthanum. The elements of the lanthanoid series all have very similar properties and applications.

The elements in the bottom row of the f-block, from Period 7, are the actinoids (pages 94 to 97), also called actinides – named after the first of them, actinium. This section of the book features only four actinoid elements, with atomic numbers up to and including 92: actinium, thorium, protactinium and uranium. These elements have few uses, apart from uranium, which is the fuel in nuclear reactors and the explosive charge in nuclear weapons. Elements with atomic numbers higher than 92 – transuranium or transuranic elements – are unstable and radioactive. The actinoids with atomic numbers higher than 92 can be found, with the other transuranium elements, on pages 153 to 157.

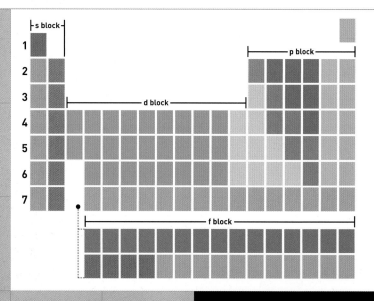

Understanding the f-block

The f-block is named after f-orbitals – just as the d-block is named after d-orbitals (see pages 38–9). The elements in the f-block have their outer electrons in f-orbitals (with the exception of lanthanum and actinium – see opposite).

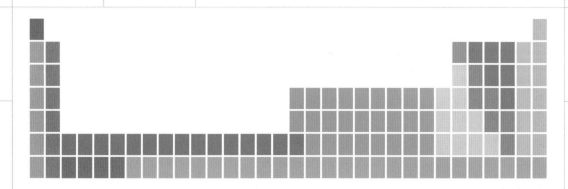

Above: The "extended" periodic table, with the f-block included, rather than separate. Here, all 118 elements discussed in this book are shown together, and the place of the lanthanoids and actinoids, in Periods 6 and 7, is clear.

While d-orbitals appear from Period 4 onwards, f-orbitals first appear in Period 6. Oddly, the f-orbitals in Period 6 are actually 4f-orbitals, so they are actually part of the fourth electron shell (n=4; see page 8). Building up the electron shells of a lanthanoid element from scratch would be like stacking shelves from the ground upwards, but leaving part of the fourth shelf empty and filling it only when you are halfway through stacking the sixth shelf. So, in Period 6, the outermost s- and p-orbitals are from the sixth shell (6s and 6p), but the f-orbitals are from the fourth (4f). As an example, here is the arrangement of Period 6 element samarium's 62 electrons, with the set for each period shown in square brackets:

$$[1s^2] \; [2s^2 \, 2p^6] \; [3s^2 \, 3p^6] \; [4s^2 \, 3d^{10} \, 4p^6] \; [5s^2 \, 4d^{10} \, 5p^6] \; \mathbf{[6s^2 \, 4f^6]}$$

A similar thing happens with d-orbitals, which are one shell, not two, behind. So, in another element from Period 6, bismuth, the (filled) f-orbitals are from the fourth shell (4f) and the outer d-orbitals are from the fifth shell (5d):

$$[1s^2] \; [2s^2 \, 2p^6] \; [3s^2 \, 3p^6] \; [4s^2 \, 3d^{10} \, 4p^6] \; [5s^2 \, 4d^{10} \, 5p^6] \; \mathbf{[6s^2 \, 4f^{14} \, 5d^{10} \, 6p^3]}$$

Despite giving their names to the lanthanoid and actinoid series, many chemists do not think the two elements lanthanum and actinium should be in the f-block at all, since they do not have their outermost electrons in f-orbitals. Instead, they suggest that their rightful place is in the gaps in Group 3. Other chemists think that it is the elements at the right-hand end of the f-block – lutetium and lawrencium – that should be in Group 3 instead (although they do have outer electrons in f-orbitals). Nevertheless, both pairs of elements are normally included in the f-block. The result is that there is one extra element in each series: there should be 14 – since there are seven f-orbitals, each with two spaces – but the lanthanoid and actinoid series each contain 15 elements.

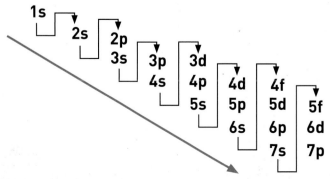

Left: The aufbau ("building up") principle. Following the arrows, it is possible to predict the electron configuration of an atom if you built it up from scratch, by successively adding electrons.

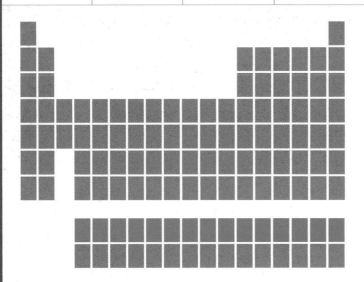

The Lanthanoids

The elements of the lanthanoid series make up the first row of the f-block of the periodic table (see page 84). In their metallic form, lanthanoids are relatively soft metals that react vigorously with oxygen if left open to the air. No lanthanoids have any known biological role. The lanthanoids, together with scandium and yttrium (see page 40), are often referred to as the rare earth metals.

The word "earths" in rare earths refers to metal oxides, and relates to the time when the existence of previously unknown metals could be inferred by identifying unfamiliar earths in minerals. The rare earth elements are not as rare as they were first thought to be, which is perhaps fortunate, for rare earths are playing ever more important roles in the modern world. They are used as catalysts, and in glassmaking, lasers, magnets and a range of useful alloys. Rare earth metals are extracted from a number of different ores, in which most exist as trace elements; the most significant rare earth ores are bastnäsite and xenotime. At present, China produces well over half of the world's supply of these important elements.

With the exception of promethium – which has no stable isotopes and does not exist naturally on Earth – the elements of the lanthanoid series are normally found closely associated in the same mineral deposits. In fact, scientists hunting for these elements in the nineteenth century relied almost entirely on two minerals, both discovered in Sweden in the eighteenth century: cerite and gadolinite. The elements cerium, lanthanum, praseodymium, neodymium, samarium and europium were all discovered within the mineral cerite. The other mineral, ytterbite (later called gadolinite) was discovered in 1787 in a quarry in the town of Ytterby, Sweden. The lanthanoids gadolinium, terbium, dysprosium, holmium, erbium, thulium, ytterbium and lutetium were all discovered in gadolinite. The discoveries of these elements, using both chemical and spectroscopic techniques, spanned more than a hundred years. Scientists were perplexed, repeatedly finding that what they thought were elements were in fact mixtures of elements. Several of the lanthanoids were not actually isolated until the 1950s, after the invention of a technology called ion-exchange.

57

La

Lanthanum

ATOMIC NUMBER: 57
ATOMIC RADIUS: 195 pm
OXIDATION STATES: +2, **+3**
ATOMIC WEIGHT: 138.91
MELTING POINT: 920°C (1,688°F)
BOILING POINT: 3,464°C (6,267°F)
DENSITY: 6.17 g/cm³
ELECTRON CONFIGURATION: [Xe] 5d¹ 6s²

The first element in the lanthanoid series, lanthanum, is used in rechargeable nickel metal hydride batteries (see page 68); an electric vehicle can contain several kilograms of this element. Lanthanum compounds were used in carbon arc lamps, but these lamps have been superseded by xenon lamps, which do not contain lanthanum. The compounds are also used in glassmaking, in the manufacture of high-quality lenses, and in phosphors.

Lanthanum was discovered in 1839, by Swedish chemist Carl Mosander. Another Swedish chemist, Jöns Jacob Berzelius, suggested the name lanthana, from the Greek word *lanthano*, which means "lying hidden" – for Mosander found lanthanum as an impurity in the mineral cerite. In 1842, Mosander went on to discover that lanthanum was actually two elements; one kept the name lanthanum, and he called the other didymium, from the Greek *didymos*, meaning "twin". Didymium itself turned out to be three elements, known today as samarium, praseodymium and neodymium.

Top: Sample of pure lanthanum.

Above: Cathode of a scanning electron microscope, from which electrons emanate. The cathode contains lanthanum hexaboride, from which electrons can be drawn very easily.

Left: Coloured X-ray of a smart phone, showing its internal structure. Several rare earth elements are used in electronic circuits, magnets and batteries for mobile electronics.

58

Ce

Cerium

Element 58, cerium, was the first of the lanthanoids to be discovered. It was first detected independently by the Swedish chemist Jöns Jacob Berzelius and German chemist Martin Klaproth, in 1803. Cerium was named after the asteroid (technically a dwarf planet) Ceres, which was first spotted in 1801. It is the most abundant of all the lanthanoids. Cerium is used as a catalyst in catalytic converters, and used in self-cleaning ovens, where it helps break up grease. It also makes up about 50 per cent of an alloy called misch metal, which is used as sparking flints in lighters.

Below: Sample of pure cerium, the most abundant rare earth element.

ATOMIC NUMBER: 58	**MELTING POINT:** 799°C (1,470°F)
ATOMIC RADIUS: 185 pm	**BOILING POINT:** 3,442°C (6,227°F)
OXIDATION STATES: +2, **+3, +4**	**DENSITY:** 6.71 g/cm^3
ATOMIC WEIGHT: 140.12	**ELECTRON CONFIGURATION:** [Xe] 4f^1 5d^1 6s^2

59

Pr

Praseodymium

In 1842, Carl Mosander discovered didymium (see lanthanum, page 87). But in 1879, French chemist Paul-Émile Lecoq de Boisbaudran found it to be a mixture of at least two elements. He isolated one (samarium), leaving the rest as didymium. In 1885, Austrian Carl Auer von Welsbach investigated didymium using spectroscopy, and found that it was two elements. He called them praseodymium, from the Greek for "green-coloured twin", and neodymium ("new twin"). Compounds of these two are used as catalysts in the petroleum industry and in vehicle catalytic converters; both are also used in iron and magnesium alloys. "Didymium" now refers to a mix of praseodymium and neodymium, used for optical coating on lenses and filters; used in safety goggles, it absorbs harsh yellow light and ultraviolet radiation.

Above: Sample of pure praseodymium.

Below: Sample of pure neodymium.

ATOMIC NUMBER: 59	**MELTING POINT:** 930°C (1,708°F)
ATOMIC RADIUS: 185 pm	**BOILING POINT:** 3,520°C (6,370°F)
OXIDATION STATES: +2, **+3**, +4	**DENSITY:** 6.78 g/cm^3
ATOMIC WEIGHT: 140.91	**ELECTRON CONFIGURATION:** [Xe] 4f^3 6s^2

60

Nd

Neodymium

For the discovery of neodymium, see praseodymium (above). Neodymium is the crucial component of an alloy used to make strong permanent magnets. Because even small neodymium magnets are strong, they are the most widely used magnets in consumer electronics applications.

ATOMIC NUMBER: 60	**MELTING POINT:** 1,020°C (1,868°F)
ATOMIC RADIUS: 185 pm	**BOILING POINT:** 3,074°C (5,565°F)
OXIDATION STATES: +2, **+3**	**DENSITY:** 7.00 g/cm^3
ATOMIC WEIGHT: 144.24	**ELECTRON CONFIGURATION:** [Xe] 4f^4 6s^2

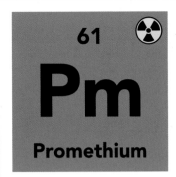

61

Pm
Promethium

Next to neodymium in the lanthanoid series is the element promethium, which nineteenth-century chemists really had no chance of finding. It is likely that less than a kilogram of promethium exists naturally on Earth; it has no stable isotopes, so most of it disintegrated a long time ago, while tiny amounts are produced from the disintegration of other radioactive elements. Promethium was first produced in 1945, in a nuclear reactor at the Oak Ridge National Laboratory in Tennessee, USA, as one of the products of the fission of uranium. The name is derived from Prometheus, a character in Greek mythology who stole fire from the god Zeus and brought it to humans. The first bulk sample of promethium metal was produced in 1963. Promethium has a few specialized applications, including its use in some nuclear batteries for spacecraft; the disintegration of the isotope promethium-147 provides the batteries' energy.

ATOMIC NUMBER: 61
ATOMIC RADIUS: 185 pm
OXIDATION STATES: +3
ATOMIC WEIGHT: (145)

MELTING POINT: 1,042°C (1,907°F)
BOILING POINT: 3,000°C (5,430°F), estimated
DENSITY AT STP: 7.22 g/cm^3
ELECTRON CONFIGURATION: [Xe] 4f^5 6s^2

62

Sm
Samarium

For the discovery of samarium, see praseodymium (opposite). French chemist Paul-Émile Lecoq de Boisbaudran, who discovered samarium, named it after the mineral from which it was first isolated, samarskite. Samarium is used as a catalyst in a process called SACRED (samarium-catalysed reductive dechlorination), which breaks down toxic chlorinated compounds. In an alloy with cobalt, samarium is used to make strong magnets, and it is one of the elements in a superconducting material – with zero electrical resistance – that could one day make power transmission and electric motors much more efficient.

ATOMIC NUMBER: 62
ATOMIC RADIUS: 185 pm
OXIDATION STATES: +2, +3
ATOMIC WEIGHT: 150.36

MELTING POINT: 1,075°C (1,967°F)
BOILING POINT: 1,794°C (3,261°F)
DENSITY AT STP: 7.54 g/cm^3
ELECTRON CONFIGURATION: [Xe] 4f^6 6s^2

Above: Sample of pure samarium.

Below: Sample of pure europium.

63

Eu
Europium

Three years after Paul-Émile Lecoq de Boisbaudran discovered samarium, he found another element – in the spectrum of light from samarium. In 1901, French chemist Eugène-Anatole Demarçay extracted it, naming it europium, for Europe. Bright red under ultraviolet radiation, it is used in fluorescent lamps.

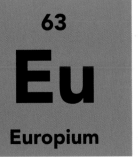

ATOMIC NUMBER: 63
ATOMIC RADIUS: 185 pm
OXIDATION STATES: +2, +3
ATOMIC WEIGHT: 151.96

MELTING POINT: 822°C (1,511°F)
BOILING POINT: 1,527°C (2,780°F)
DENSITY AT STP: 5.25 g/cm^3
ELECTRON CONFIGURATION: [Xe] 4f^7 6s^2

64
Gd
Gadolinium

ATOMIC NUMBER: 64
ATOMIC RADIUS: 180 pm
OXIDATION STATES: +1, +2, **+3**
ATOMIC WEIGHT: 157.25
MELTING POINT: 1,313°C (2,395°F)
BOILING POINT: 3,272°C (5,921°F)
DENSITY: 7.87 g/cm³
ELECTRON CONFIGURATION: [Xe] 4f⁷ 5d¹ 6s²

Using spectroscopy, Swiss Chemist Jean de Marignac detected the signature of an unknown element in a sample of terbia (the oxide of terbium, below), in 1880. French chemist Paul-Émile Lecoq de Boisbaudran named the element after he became the first to isolate it, in 1886. The name is derived from the mineral gadolinite. Like samarium, gadolinium is very good at absorbing neutrons produced in nuclear reactions; as a result, it is used in shielding and control rods in nuclear reactors. Gadolinium compounds are injected into some patients undergoing MRI (magnetic resonance imaging) scans; their presence in a patient's tissues help to improve the clarity of the images produced.

Top right: Sample of pure gadolinium.

Right: : Coloured magnetic resonance imaging (MRI) scan showing a tumour (red) in one kidney. Gadolinium nanoparticles injected into the patient's blood accumulate in the tumour, enabling it to be detected easily.

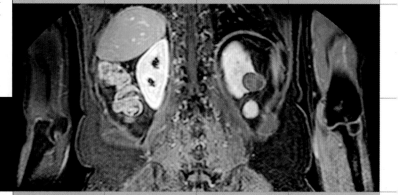

65
Tb
Terbium

In 1843, Swedish chemist Carl Mosander found two unknown elements in yttria (gadolinite). Mosander called the elements terbium and erbium, after the mineral in which they were discovered, ytterbia – which in turn was named after the Swedish village of Ytterby. Due to confusion and disagreement in the 1860s, the names of these two elements became swapped. In 1886, Jean de Marignac discovered that terbium (Mosander's erbium) was actually two elements – gadolinium and modern-day terbium. Today, the most important use of terbium is in yellow-green phosphors used in fluorescent lamps, but terbium alloys are also used in some solid state electronics.

Below: Sample of pure terbium.

ATOMIC NUMBER: 65
ATOMIC RADIUS: 175 pm
OXIDATION STATES: +1, **+3**, +4
ATOMIC WEIGHT: 158.93

MELTING POINT: 1,356°C (2,473°F)
BOILING POINT: 3,227°C (5,840°F)
DENSITY: 8.27 g/cm³
ELECTRON CONFIGURATION: [Xe] 4f⁹ 6s²

66
Dy
Dysprosium

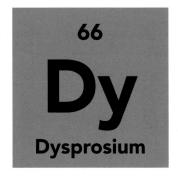

In 1886, French chemist Paul-Émile Lecoq de Boisbaudran discovered the oxide of an unknown element in a sample of the oxide of another lanthanoid, holmium (see below). De Boisbaudran named the element dysprosium, after the Greek word *dysprositos*, meaning "hard to obtain"; it was only after many attempts that he managed to produce a small sample of it. Dysprosium has some applications in lasers and lighting, and it is increasingly being used together with neodymium in strong permanently magnetic alloys, particularly in motors for electric cars.

Above: Sample of pure dysprosium.

ATOMIC NUMBER: 66	**MELTING POINT:** 1,412°C (2,573°F)
ATOMIC RADIUS: 175 pm	**BOILING POINT:** 2,567°C (4,653°F)
OXIDATION STATES: +2, **+3**	**DENSITY:** 8.53 g/cm^3
ATOMIC WEIGHT: 162.50	**ELECTRON CONFIGURATION:** [Xe] 4f^{10} 6s^2

67
Ho
Holmium

Swedish chemist Per Teodor Cleve discovered the oxides of three elements – thulium, erbium and holmium – in what had been considered just one: erbium. The name holmium is from the Latin name for Stockholm, Holmia. Today, holmium's primary use is medical and military lasers. Holmium oxide give cubic zirconia jewellery a yellow colour

ATOMIC NUMBER: 67	**MELTING POINT:** 1,474°C (2,685°F)
ATOMIC RADIUS: 175 pm	**BOILING POINT:** 2,700°C (4,885°F)
OXIDATION STATES: **+3**	**DENSITY:** 8.80 g/cm^3
ATOMIC WEIGHT: 164.93	**ELECTRON CONFIGURATION:** [Xe] 4f^{11} 6s^2

68
Er
Erbium

The name erbium was originally used for the element terbium, after Carl Mosander detected these elements in the mineral ytterbite (see terbium, opposite). The names switched during the 1860s, and, to confuse matters still more, each of these elements was actually a mixture of two or more elements. In 1879, Per Teodor Cleve found that what chemists had thought was pure erbium oxide also contained the oxides of two other elements (thulium and holmium). Like holmium, erbium is used in lasers and its oxide is used in tiny amounts in cubic zirconium used in jewellery, in which it produces a pink colour. Erbium is also used as a dopant in fibre-optic cables, where it reduces the loss of signal strength; and a small amount of erbium added to vanadium makes an alloy that is easier to work and softer to the touch.

ATOMIC NUMBER: 68	**MELTING POINT:** 1,529°C (2,784°F)
ATOMIC RADIUS: 175 pm	**BOILING POINT:** 2,868°C (5,194°F)
OXIDATION STATES: **+3**	**DENSITY:** 9.04 g/cm^3
ATOMIC WEIGHT: 167.26	**ELECTRON CONFIGURATION:** [Xe] 4f^{12} 6s^2

69

Tm

Thulium

ATOMIC NUMBER: 69

ATOMIC RADIUS: 175 pm

OXIDATION STATES: +2, **+3**

ATOMIC WEIGHT: 168.93

MELTING POINT: 1,545°C (2,813°F)

BOILING POINT: 1,950°C (3,542°F)

DENSITY: 9.33 g/cm³

ELECTRON CONFIGURATION: [Xe] $4f^{13}$ $6s^2$

Thulium is probably the least abundant of all the lanthanoids (except for unstable promethium) – although it is still around a hundred times more abundant than gold. It was one of the elements Per Teodor Cleve discovered, as thulium oxide, in a sample of erbium oxide in 1879 (see erbium, page 91). Cleve named the new element after the ancient word for a mysterious, northerly part of Scandinavia, Thule – although he misspelt it, with a double "l," and thought the word referred to the whole of Scandinavia. Like many of the lanthanoids, thulium is used in the crystals that do the lasing in certain types of laser – and, like samarium, may one day be used in superconducting materials.

Above: Sample of pure thulium.

70

Yb

Ytterbium

ATOMIC NUMBER: 70

ATOMIC RADIUS: 175 pm

OXIDATION STATES: +2, **+3**

ATOMIC WEIGHT: 173.04

MELTING POINT: 824°C (1,515°F)

BOILING POINT: 1,196°C (2,185°F)

DENSITY: 6.95 g/cm³

ELECTRON CONFIGURATION: [Xe] $4f^{14}$ $6s^2$

In 1878, Swiss chemist Jean de Marignac found the signature of a new element in his spectroscopic analysis of terbium – which, confusingly, had been known as erbium until the 1860s (see erbium, page 91). Ytterbium was the fourth element to be named after the Swedish village of Ytterby, the others being terbium, erbium and yttrium. Swedish chemist Lars Nilson discovered that Marignac's ytterbium was in fact two elements; one was the Group 3 element scandium (see page 40), which is a rare earth element but not a lanthanoid; the other retained the name ytterbium. But in 1907, French chemist Georges Urbain showed that even this newly

Left: Sample of pure ytterbium.

purified ytterbium was a mixture, comprising a new element he called lutecium (now lutetium) and what he then called neoytterbium. The name ytterbium was only formally accepted in 1925, and a pure sample of the element only obtained in the 1950s. Today, ytterbium is used in some stainless steels, in solar cells and lasers. It is also used, together with erbium, in anti-forgery inks on bank notes: ytterbium sensitizes erbium to glow red or green under infrared light.

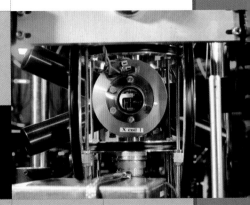

Right: Ytterbium optical clock at the National Physical Laboratory, UK. Atomic clocks based on visible radiation from ytterbium ions are thousands of times more accurate than caesium-based atomic clocks, and could be the standard timekeepers of the future.

71

Lu

Lutetium

ATOMIC NUMBER: 71

ATOMIC RADIUS: 175 pm

OXIDATION STATES: **+3**

ATOMIC WEIGHT: 174.97

MELTING POINT: 1,663°C (3,025°F)

BOILING POINT: 3,400°C (6,150°F)

DENSITY: 9.84 g/cm3

ELECTRON CONFIGURATION: [Xe] $4f^{14} 5d^1 6s^2$

The last of the lanthanoid series, and the last stable lanthanoid to be identified, lutetium was discovered by French chemist Georges Urbain, in 1907 (see ytterbium, above). Austrian mineralogist Baron Carl Auer von Welsbach and American chemist Charles James independently discovered the element in the same year. Urbain proposed the name lutecium, with a "c", from the Latin name for Paris, *Lutetia*. The spelling was changed in 1949. Lutetium is one of the least abundant of the lanthanoids, but still has some applications. For example, it is used as a catalyst in the chemical industry, in particular in "cracking" hydrocarbons in oil refineries, and in sensors in medical scanners.

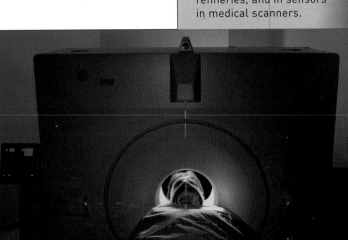

Above: Sample of pure lutetium.

Left: A patient being scanned in a positron emission tomography (PET) scanner. A radioactive substance injected into the patient's body produces positrons; these are detected by a sensor consisting of a compound of lutetium doped with another rare earth, cerium.

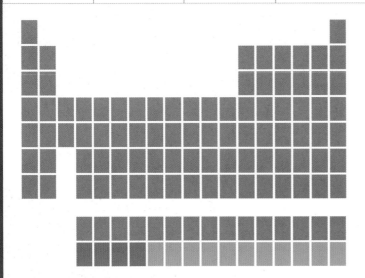

The Actinoids

The 15 elements of the actinoid series are heavy elements, with between 89 and 103 protons in the nuclei of their atoms, in Period 7 of the periodic table. They make up the second row of the table's f-block (see page 84).

All the actinoids are radioactive – their nuclei spontaneously disintegrate, releasing subatomic particles and transmuting into different elements. However, thorium and uranium have isotopes with half-lives long enough that quantities of these elements created in supernovas billions of years ago still persist here on Earth. Additionally, protactinium, neptunium and plutonium are found naturally in tiny amounts in uranium ores, their existence the result of the disintegration of uranium nuclei. Those elements of the actinoid series with atomic numbers higher than uranium's were only discovered as a result of their being synthesized in nuclear reactors since the middle of the twentieth century. These elements feature with the other transuranium elements, on pages 153–7.

In early versions of the periodic table – before most of the actinoids had been discovered or synthesized – uranium was placed beneath tungsten, and thorium was placed beneath hafnium. After a few more actinoids had been created, in nuclear laboratories in the 1940s, chemists and physicists realized that these elements wouldn't fit neatly into the grouped structure of the periodic table as it stood. And so, in 1944, American physicist Glenn Seaborg proposed the actinide hypothesis. Seaborg suggested that just as the elements from lanthanum to lutetium form a series of elements that break away from the standard structure of the periodic table (see page 84), so actinium, thorium, protactinium, uranium and the new heavy elements might also form a series.

By far the most important element in the actinoid series is uranium. One isotope of uranium, uranium-235, is the main source of fuel in nuclear reactors. Most of the other actinoids – even the synthetic ones – have applications, albeit very limited.

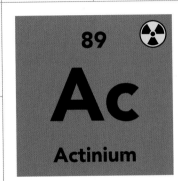

89

Ac
Actinium

Actinium was discovered in 1899, by French chemist André-Louis Debierne, in the uranium ore from which Pierre and Marie Curie discovered radium a year earlier. Debierne gave the new element the name "actinium" from the Greek word for ray, aktinos; the Latin equivalent, radius, had been used to name radium. Actinium only has a few, specialized, applications because of its rarity and radioactivity.

ATOMIC NUMBER: 89
ATOMIC RADIUS: 195 pm
OXIDATION STATES: +3
ATOMIC WEIGHT: (227)

MELTING POINT: 1,050°C (1,922°F)
BOILING POINT: 3,200°C (5,790°F)
DENSITY: 10.06 g/cm³
ELECTRON CONFIGURATION: [Rn] 6d¹ 7s²

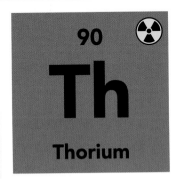

90

Th
Thorium

Thorium and uranium are the only actinoids whose atoms have survived in significant numbers since they were created in supernovas billions of years ago. The most stable isotope, thorium-227, has a half-life of around 14 billion years, which means that only around 20 per cent of the thorium atoms present when Earth was created around 4.6 billion years ago have now decayed. Thorium was discovered by Swedish chemist Jöns Jacob Berzelius, in 1829, in a mineral very similar to ytterbite (see page 86). Berzelius chose the name "thorium" – derived from the Norse god of war, Thor. (He had first used this name in 1815, when he thought he had discovered a new element; however, that turned out to be yttrium, which was already known.) Nearly pure thorium metal was first prepared in 1914. Thorium oxide has a very high melting point, and is used in producing heat-resistant ceramics and welding electrodes. Thorium itself is sometimes used in high-performance alloys with magnesium.

ATOMIC NUMBER: 90
ATOMIC RADIUS: 180 pm
OXIDATION STATES: +2, +3, **+4**
ATOMIC WEIGHT: 232.04

MELTING POINT: 1,750°C (3,190°F)
BOILING POINT: 4,790°C (8,650°F)
DENSITY: 11.73 g/cm³
ELECTRON CONFIGURATION: [Rn] 6d² 7s²

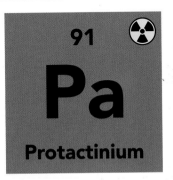

91

Pa
Protactinium

Tiny amounts of protactinium exist naturally, the result of the decay of uranium-235 in uranium ores. Dmitri Mendeleev predicted the existence of an element below tantalum – "eka-tantalum" – in 1871. It was discovered in 1913, as one of the elements in the "decay chain" of uranium, by Polish-American physicist Kazimierz Fajans and German physicist Oswald Göhring. These two physicists initially chose the name "brevium" – brevis is Latin for "short-lived". The name was later changed to "protoactinium" (protos is Greek for "before") – because the element transmutes into actinium when it decays – and shortened to its present form in 1949. Protactinium only has a few, specialized, applications because of its rarity and radioactivity.

ATOMIC NUMBER: 91
ATOMIC RADIUS: 180 pm
OXIDATION STATES: +3, +4, **+5**
ATOMIC WEIGHT: 231.04

MELTING POINT: 1,572°C (2,862°F)
BOILING POINT: 4,000°C (7,230°F), approximate
DENSITY: 15.37 g/cm³
ELECTRON CONFIGURATION: [Rn] 5f² 6d¹ 7s²

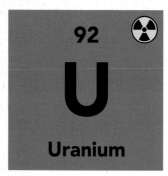

92

U

Uranium

ATOMIC NUMBER: 92

ATOMIC RADIUS: 175 pm

OXIDATION STATES: +3, +4, +5, **+6**

ATOMIC WEIGHT: 238.03

MELTING POINT: 1,134°C (2,037°F)

BOILING POINT: 4,130°C (7,470°F)

DENSITY: 19.05 g/cm³

ELECTRON CONFIGURATION: [Rn] $5f^3$ $6d^1$ $7s^2$

People have been mining uranium ores for hundreds of years; long before uranium was discovered as a chemical element, the uranium ore pitchblende was used as an additive in glassmaking that turned the glass yellow. In 1789, German chemist Martin Klaproth realized that pitchblende contained an unknown element, which he named after the planet Uranus (discovered in 1781). Klaproth prepared a small amount of black powder that he thought was pure uranium, but was actually uranium(IV) oxide (UO_2). French chemist Eugène-Melchior Péligot produced the first sample of uranium metal in 1841. It was by studying pitchblende that French physicist Antoine Henri Becquerel discovered radioactivity, in 1896. Becquerel's doctoral student, the Polish physicist Marie Curie, coined the term "radioactivity" in 1898 after studying the phenomenon. Curie studied the invisible rays emitted by uranium, and showed that they were being produced by the uranium atoms themselves, not as a result of chemical reactions, or of heat or light.

Three isotopes of uranium exist in nature, all of them radioactive: uranium-234, uranium-235 and uranium-238. Of these, uranium-238 has the longest half-life: more than 4 billion years. Uranium-235, which makes up less than 1 per cent of naturally occurring uranium, is fissile: when neutrons hit the nuclei of atoms of this isotope, the nuclei split in two, releasing energy and more neutrons. The released neutrons can then go on to hit other uranium-235 nuclei, initiating a nuclear fission chain reaction and releasing large amounts of energy, ultimately as intense heat. If there is enough uranium-235 present – above a critical mass – the chain reaction can become sustained. The enormous amount of heat produced in such sustained reactions is used in nuclear power stations to generate electricity, and is the source of the destructive force of nuclear weapons.

Nuclear fission was discovered in 1938, after experiments involving the bombardment of uranium with neutrons had the puzzling result that a much lighter element, barium, was produced. Physicists had expected that adding neutrons would increase the mass of the nuclei, or at most cause nuclei to shed an alpha particle (see pages 9–10). Austrian physicist Lise Meitner and German chemist Otto Hahn were the first to work out fully what was going on.

Above: Nuclear fuel pins. Each pin is a metal tube containing pellets of uranium and plutonium oxides. Hundreds of these pins assembled together form a unit that sits in the core of a nuclear reactor.

Above right: Uraninite, a radioactive mineral, mostly consisting of uranium oxide.

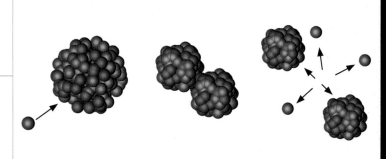

Left: Sequence showing neutron-induced nuclear fission. A neutron hits a nucleus of uranium-235, which becomes a very unstable nucleus of uranium-236. The nucleus splits into two lighter daughter nuclei, and releases three neutrons, which can initiate fission in other nuclei, starting a chain reaction.

Below left: French nuclear submarine *Saphir*, which is powered by a small nuclear reactor.

Below: Sequence showing the first 0.1 seconds of the Trinity test, the detonation of the world's first atomic bomb, on 16 July 1945, in New Mexico, USA.

The fuel in most nuclear power stations is uranium(IV) oxide, which has a higher melting temperature than the metal itself and cannot react with air. The uranium used is enriched – the proportion of uranium-235 has been increased, so that a chain reaction is possible. The enrichment most commonly takes place in a gas centrifuge: uranium(VI) fluoride (UF_6) gas is pumped into a rapidly rotating cylinder, in which the slightly heavier uranium-238 fluoride molecules end up closer to the cylinder's wall, while the lighter uranium-235 fluoride molecules remain closer to the centre. This process is repeated many times, increasing the proportion of uranium-235 to between 3 per cent and 20 per cent for use in nuclear reactors and up to 90 per cent for nuclear weapons.

The waste product of the enrichment process is depleted uranium, so called because it has a lower-than-normal proportion of uranium-235. This depleted uranium is still radioactive, but less so than natural uranium. It is denser than lead, and it is alloyed with other metals and used to make military armour plating and shells and shielding for X-ray machines. Spent nuclear fuel is also called depleted uranium, because most of the uranium-235 that was present has undergone fission. But it is more radioactive than the other type of depleted uranium, because it contains some transuranium elements created inside the reactor.

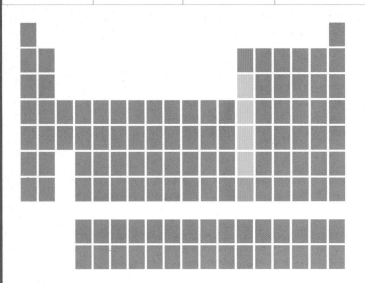

5
B
Boron

13
Al
Aluminium

31
Ga
Gallium

49
In
Indium

81
Tl
Thallium

113
Uut
Ununtrium

The Boron Group

The best-known element of Group 13 is aluminium – one of the few we commonly encounter in elemental form (although aluminium foil and cans are only about 95 per cent aluminium). Boron, gallium, indium and thallium are less well known, but are not particularly rare and have a variety of applications. This group also includes element 113, temporarily named "ununtrium" (1-1-3-ium) – see transuranium elements on pages 153–7.

The atoms of all these elements have three electrons in their (incomplete) outer shell – two in an s-orbital and one in a p-orbital (see pages 8–9); they all have the electron configuration $s^2 p^1$. Since it is the outer, or valence, electrons that take part in chemical reactions and determine an element's properties, you might expect these elements to be very similar. But boron is a non-metal (actually a metalloid; see opposite), and forms only covalent bonds (see pages 11–12); aluminium and gallium are metallic, but with some non-metal characteristics – and they form both covalent and ionic bonds; indium and thallium are true metals, and form only ionic bonds.

This trend down the group is explained by the sizes of the atoms: boron atoms are small, having only two electron shells, while thallium atoms are much bigger, with seven electron shells. Being further away from the nucleus, the outer electrons of indium and thallium are more easily lost, thus forming ions. This trend, from non-metallic to metallic, is repeated in Groups 14, 15 and 16, while Groups 1 to 12 contain only metals and Groups 17 and 18 contain only non-metals. Group 13 is the first group in the p-block of the periodic table; the atoms of these elements all have their outer electrons in p-orbitals.

5

B

Boron

ATOMIC NUMBER: 5

ATOMIC RADIUS: 85 pm

OXIDATION STATES: +1, +2, **+3**

ATOMIC WEIGHT: 10.81

MELTING POINT: 2,079°C (3,774°F)

BOILING POINT: 4,000°C (7,232°F)

DENSITY: 2.47 g/cm³

ELECTRONIC COFIGURATION: [He] 2s² 2p¹

The lightest of the Group 13 elements, boron, is classified as a metalloid – with some characteristics of metals and some of non-metals. Most elements whose atoms have only three electrons in their outer shell are metals, because it is easy to lose those electrons, leaving a stable, complete shell of electrons – and resulting in metallic bonding, in which the electrons are free to wander between the metal atoms, and ionic bonding in compounds with non-metals. But boron atoms are small, and hold on to their electrons more tightly. As a result, boron does not take part in metallic bonding and forms only covalent bonds (see pages 11–12).

Boron is much more rare than most of the other light elements; it is the thirty-eighth most abundant element in Earth's crust. Most elements up to atomic number 26 (iron) were made inside stars – but not boron. And although some boron was made shortly after the Big Bang, that was destroyed inside stars. Instead, boron nuclei are created when fast-moving protons (cosmic rays) strike other atomic nuclei – a process called spallation.

Despite its relative rarity, boron is found in more than a hundred minerals. The most important mineral of boron is borax, which has many uses itself – most notably, in detergents and cosmetics. Borax has been used and mined for hundreds of years; the name "boron" derives from the Ancient Arabic and Persian names for the mineral,

Above: Crystals of borax, which is hydrated sodium borate (Na₂B₄O₇.10H₂O). Borax forms during the evaporation of salty lakes.

Right: Sample of the metalloid element boron.

Above: Soldiers wearing bulletproof vests. The protective plates incorporated into the vest are made of the extremely hard compound boron carbide.

Right: Pyrex® jug, made from borosilicate glass, which has a low coefficient of expansion. This means that it is less likely than ordinary glass to shatter with sudden changes of temperature, such as contact with boiling water.

buraq and *burah*. Boron itself was first isolated from borax by several chemists in 1808 – most notably the English chemist Humphry Davy, who prepared it using electrolysis.

Glass used for making cookware and laboratory glassware is normally borosilicate glass, which is made with boron(III) oxide (B_2O_3). Borosilicate glass is also used to make the glass fibres for fibreglass, and in some optical fibres. Boron itself is used to dope pure silicon in the manufacture of computer chips and solar cells; the three outer electrons of boron contrasted with the four outer electrons of silicon leave a net positively charged "hole", which can migrate through the material.

13

Al

Aluminium

ATOMIC NUMBER: 13

ATOMIC RADIUS: 125 pm

OXIDATION STATES: +1, **+3**

ATOMIC WEIGHT: 26.98

MELTING POINT: 660°C (1,220°F)

BOILING POINT: 2,519°C (4,566°F)

DENSITY: 2.70 g/cm³

ELECTRON CONFIGURATION: [Ne] $3s^2\ 3p^1$

The third most abundant element – and the most abundant metal – in Earth's crust, aluminium is plentiful and has myriad uses. It is light and strong, and is used as a structural metal in buildings and vehicles. When exposed to the air, pure aluminium quickly reacts with oxygen, forming a thin layer of aluminium(III) oxide (Al_2O_3), which prevents further reaction. And so, although aluminium is a reactive element, aluminium metal is resistant to corrosion. Structural components made of aluminium are usually cast – but aluminium can also be rolled into a foil, drawn out into wires and made into a powder. Aluminium powder is used in theatrical flash powder and in fireworks. The powder is also used in the manufacture of glass mirrors; it is heated in a vacuum to create a vapour of aluminium, which deposits a thin, even coating on to the glass. Aluminium is a good conductor of electricity; for this reason, and because it is relatively cheap and lightweight, most overhead and underground power transmission cables are made of it.

Right: Globule of freshly re-melted pure aluminium. The surface of a sample of this metal retains its brilliant lustre despite the rapid formation of an oxide layer.

Several aluminium compounds are also used in large quantities – most notably aluminium(III) oxide, which is used as a catalyst and an abrasive. Aluminium(III) hydroxide ($Al(OH)_3$) is used as an antacid to ease indigestion, and aluminium chlorohydrate ($Al_2Cl(OH)_5$) is the most common active ingredient in antiperspirants. Another

compound, aluminium(III) sulfate ($Al_2(SO_4)_3$), is used as part of the purification process for water and treatment of sewage. Most of these aluminium compounds are obtained or prepared from bauxite, a mineral that is also the most common ore from which aluminium is extracted.

The name "aluminium" is derived from alum, a mineral containing the compound potassium aluminium sulfate ($K_2Al_2(SO_4)_4$). In 1787, French chemist Antoine Lavoisier realized that a component of alum, which was known as "alumine", contained an unknown metal; in 1808, Humphry Davy suggested that the new element be called "alumium", but this quickly changed to aluminum and then aluminium. (Both versions, "-um" and "-ium", are equally valid.) German chemist Friedrich Wöhler prepared a fairly pure sample of aluminium in 1827. Although metallurgists recognized the metal's great potential, for many years there was no way to extract aluminium cheaply; as a result, it was worth more than gold and silver. In 1886, French inventor Paul Héroult and American chemist Charles Martin Hall both independently devised the same method of extracting aluminium from aluminium oxide, using electricity. The price of the metal plummeted, and aluminium quickly became widely available; the Hall-Héroult method is still the basis of present-day aluminium production. Today, around 50 million tonnes of aluminium metal are produced each year (a quarter of it recycled); this is nearly three times as much as copper and second only to iron.

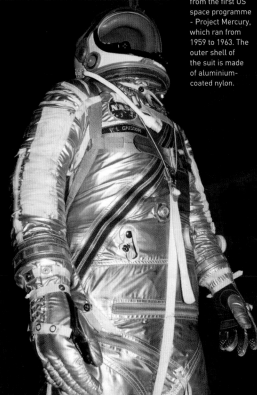

31

Ga

Gallium

ATOMIC NUMBER: 31

ATOMIC RADIUS: 130 pm

OXIDATION STATES: +1, +2, **+3**

ATOMIC WEIGHT: 69.72

MELTING POINT: 30°C (86°F)

BOILING POINT: 2,205°C (4,000°F)

DENSITY: 5.91 g/cm³

ELECTRON CONFIGURATION: [Ar] 3d¹⁰ 4s² 4p¹

Pure gallium is a shiny metal with a low melting point – just 30°C. A sample of gallium will melt if it is held in the hand. (Although the resulting liquid metal will stain the skin slightly, it is non-toxic.) An alloy called Galinstan®, containing around two-thirds gallium (along with indium and tin), melts at -19°C, and is increasingly used as a substitute for toxic mercury in thermometers.

There are few applications of metallic gallium. Most of the demand for the metal is from the semiconductor industry, either for doping silicon to produce transistors, or for producing the compounds gallium(III) arsenide (GaAs) and gallium(III) nitride (GaN). Some integrated circuits (chips) – particularly those used in telecommunications – are made using these compounds rather than silicon. The light-emitting crystals inside LEDs (light-emitting diodes) and laser diodes contain mixtures of several gallium compounds. Gallium arsenide is used in red and infrared LEDs, and gallium arsenic phosphide in yellow LEDs, for example. Blue and ultraviolet LEDs and laser diodes contain gallium nitride – and it was the invention of these components that made possible high-density Blu-ray™ optical storage discs. Gallium arsenide is also used to make efficient solar cells, although their expense restricts their use to high-end applications such as providing power for satellites and space probes.

Scientists have found that gallium nitride is biocompatible: in other words, the human body will not reject it. This discovery might lead to safer and more effective electrodes for stimulating neurones in the brain as part of treatment for neurological disorders such as Alzheimer's disease, or making tiny electronic chips that could be implanted to monitor various tissues. The radioactive isotope gallium-67 is used as part of a gallium scan: an imaging technique that can pinpoint tumours or areas of infection in a person's body.

Gallium was one of the elements whose existence Dmitri Mendeleev predicted in 1871, to fill the gaps in his periodic table; Mendeleev called it *eka-aluminium* ("below aluminium"). It was only four years before gallium was discovered, by French chemist Paul-Émile Lecoq de Boisbaudran. At first, de Boisbaudran identified the missing element using spectroscopy, and later the same year he extracted metallic gallium using electrolysis. The name "gallium" is derived from the Latin word *Gallia*, which refers to the ancient land of Gaul, the precursor to modern-day France – de Boisbaudran's home country.

Today, about 200 tonnes of gallium metal are produced worldwide each year, mostly as a by-product of aluminium extraction. China, Germany and Kazakhstan are the biggest producers, and much of the gallium produced is recovered by recycling semiconductors.

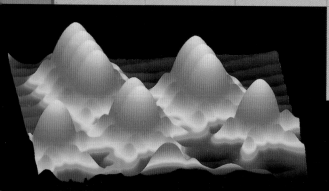

Above: A sample of gallium metal. The element's melting point is low enough that this metal will melt when held for a few minutes in a person's hand.

Left: Image produced by a scanning tunnelling microscope, showing individual atoms at the surface of gallium manganese arsenide. This compound is the most promising material for use in "spintronics", an emerging form of electronics with dramatically increased processing power.

49

In

Indium

ATOMIC NUMBER: 49
ATOMIC RADIUS: 155 pm
OXIDATION STATES: +1, +2, **+3**
ATOMIC WEIGHT: 114.82
MELTING POINT: 157°C (314°F)
BOILING POINT: 2,075°C (3,770°F)
DENSITY: 7.30 g/cm³
ELECTRON CONFIGURATION: [Kr] 4d¹⁰ 5s² 5p¹

The last two elements of Group 13 are both named after the colours of bright lines in the spectrum of the light they produce when their compounds are heated in a flame. Element 49, indium, was named after the colour indigo. It was discovered in 1863, by German chemists Ferdinand Reich and Hieronymous Theodor Richter, during a spectroscopic study of minerals thought to contain thallium (see below). The indigo-blue line they observed did not match any of the known elements at the time, and they realized they had discovered a new element.

In its pure form, indium is a bright silvery metal. Only a few hundred tonnes of indium metal are produced each year, mostly as a by-product of the extraction of zinc. The most important application of indium is in the production of electronic displays. A solid mixture of indium(III) and tin(IV) oxides ($[In_2O_3]$ and SnO_2]) has an unusual combination of properties: it is both transparent and a good conductor of electricity. Most flat panel electronic displays need electrodes at the front as well as the back; the electrodes at the front have to be transparent, or else the image would be obscured. Another compound, indium(III) nitride, is used along with gallium(III) nitride in LEDs (see opposite).

Above: A strip of pure indium foil.

81

Tl

Thallium

ATOMIC NUMBER: 81
ATOMIC RADIUS: 190 pm
OXIDATION STATES: +1, +3
ATOMIC WEIGHT: 204.38
MELTING POINT: 303°C (577°F)
BOILING POINT: 1,473°C (2,683°F)
DENSITY: 11.86 g/cm³
ELECTRON CONFIGURATION: [Xe] 4f¹⁴ 5d¹⁰ 6s² 6p¹

The heaviest element in Group 13 with a stable isotope is thallium, named after the bright green line observed in its spectrum. The name is derived from the Greek word *thallos*, which means "green shoot". Thallium was discovered in 1861, by two researchers working independently: the English physicist William Crookes and the French chemist Claude-Auguste Lamy. The element has a few rather specialized applications, mostly in electronics, and only a few tonnes of thallium metal are produced each year, mostly as a by-product of copper and zinc smelting.

Thallium is highly toxic to humans and other animals. Thallium ions in solution have a similar size to ions of potassium, which are essential to many vital processes. Inside the body, thallium is taken up through the same mechanisms as potassium – and it accumulates in particular in the brain, the kidneys and the muscle of the heart. In the kidneys, thallium is retained rather than being excreted, because that is what happens to potassium. The unpleasant symptoms of thallium poisoning include vomiting, pain, restlessness, hallucinations and loss of hair – and ultimately an untimely death.

Above: Pure thallium metal.

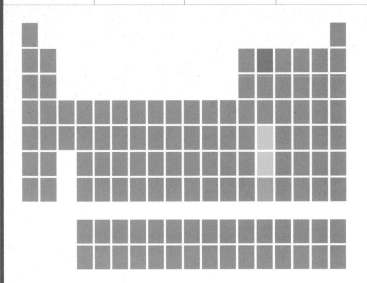

6 **C** Carbon	
14 **Si** Silicon	
32 **Ge** Germanium	
50 **Sn** Tin	
82 **Pb** Lead	
114 ☢ **Fl** Flerovium	

The Carbon Group

In their elemental state, the elements of Group 14 vary considerably: carbon is a black (or transparent, as diamond) non-metal; silicon and germanium are semiconducting metalloids; tin and lead are lustrous silvery metals. They form a wide variety of different compounds, and have an enormous range of applications. This group also includes element 114, flerovium (see the transuranium elements, on pages 153–7).

All the elements of Group 14 have atoms with four electrons in their outer shell (see pages 8–9). Each element's electron configuration ends with $s^2 p^2$. The part-filled outer electron shell leads to some interesting chemical properties – particularly for carbon. There is a tendency for the s- and p-orbitals to merge, forming sp^3 and sp^2 hybrid orbitals, which share the electrons more evenly around the nucleus and provide for some interesting opportunities for bonding with other atoms. Carbon's ability to form hybrid orbitals makes it a very versatile element – there are more carbon-containing compounds than all other compounds, and carbon is the basis of all life on Earth. If you are looking for elements around you in your everyday life, carbon is ubiquitous – forming the basis of wood, paper, plastics, oil and gas, for example.

All the elements of Group 14 have available conduction bands – energy levels in which electrons are free to move between atoms, delocalized in a kind of electron "soup". This property is characteristic of metals (see pages 10–11). The further down the group, the more easily electrons can move into the conduction band. So carbon is not generally a good conductor of electricity – although in the form of graphite, it can conduct fairly well. Silicon and germanium are classic semiconductors: they can be made to conduct if their electrons are given extra energy, from light or heat. Tin and lead are true metals: some of their electrons permanently reside in the conduction band.

6
C
Carbon

ATOMIC NUMBER: 6

ATOMIC RADIUS: 70 pm

OXIDATION STATES: **-4**, -3, -2, -1, +1, +2, +3, **+4**

ATOMIC WEIGHT: 12.01

MELTING/BOILING POINT: Sublimes at around 3,600 °C (6,500 °F)

DENSITY: 3.52 g/cm^3 (diamond), 2.27 g/cm^3 (graphite)

ELECTRONIC CONFIGURATION: [He] 2s^2 2p^2

Above right: A diamond – one form of pure carbon. This example is a "brilliant": a diamond that has been cut many times, to give it many reflective facets.

Below left: Concept for a "smart pill" drug delivery system. The pill features a nanoscale radio antenna made from carbon nanotubes, and a carbon nanotube motor to disperse the medication.

Below right: Graphite pencil – the "lead" is made from graphite mixed with clay.

Carbon is the fourth most abundant element in the Universe and fifteenth most abundant in Earth's crust – where it occurs primarily in carbonate minerals, such as limestone ($CaCO_3$), and as fossil fuels, such as coal. Carbon is wonderfully diverse even when pure, and combined with other elements it is even more versatile. Another notable feature is that there is no standard melting or boiling point given for carbon, because at normal pressures, it sublimes – turns from a solid directly into a gas.

Pure carbon has several distinct forms, or allotropes. (Several other elements have allotropes, too – see tin and phosphorus, for example.) The main allotropes of carbon are diamond, graphite, graphene, amorphous carbon and a range of substances called fullerenes.

Diamonds are formed under high pressure and at high temperature, about 150 kilometres deep in the upper mantle – a treacle-like layer of semi-molten rock beneath Earth's crust – and are brought to the surface by eruptions of magma. In diamond, each carbon atom is bonded to four others by covalent bonds (see pages 11–12), forming repeating tetrahedrons. All the electrons are involved in these bonds, and this makes diamond incredibly hard and strong – and it is because no electrons are free to absorb light passing through that diamond is transparent.

In graphite, each carbon atom is joined to three others, forming flat planes. There is one electron per atom left over, and this becomes delocalized and free from its atom; the free electrons mean that graphite is a conductor (and opaque). It is used as the contacts in some electric motors and in zinc-carbon batteries. Graphite's planes are held together very loosely, and they can slip past each other and separate. This makes graphite a good lubricant – and is also why it is used in pencils (normally mixed with baked clay). The name "graphite" dates from 1789, and comes from its use in pencils; *graphein* is the Greek word for "to write". Graphite can be made into tiny fibres much thinner than a human hair. Mixed with polymers (also carbon-based compounds), these fibres help make tough carbon fibre-reinforced plastics, which have many uses, from consumer goods such as tennis rackets to applications in industry and aerospace.

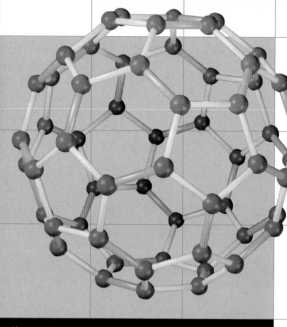

Above: In this computer model of the molecular structure of buckminsterfullerene (C_{60}), carbon atoms appear as grey spheres, double bonds are red and single bonds are cream.

Left: Manufacture of polythene (polyethene) sheets for making bags. Each polythene molecule typically consists of around 10,000 ethene molecules joined together – a total of about 20,000 carbon atoms and 40,000 hydrogen atoms.

Less well-known than diamond or graphite is a class of substances called fullerenes. These are molecules consisting only of carbon atoms, joined in hexagonal and pentagonal rings. The first fullerene to be discovered, in 1985, was buckminsterfullerene, which occurs as spherical molecules containing 60 carbon atoms. The molecule's name derives from American architect Richard Buckminster Fuller, who designed geodesic domes that have a structure similar to the molecule's atomic arrangement. Research into these spherical molecules led to the discovery of carbon nanotubes – basically, sheets of graphite rolled into tiny cylinders – along with methods of creating them. Carbon nanotubes are typically a few nanometres (10^{-9}m) wide but can be more than a millimetre long. They are remarkably strong, and have interesting electrical properties. Another carbon allotrope to come to the fore as a result of the rise of fullerenes is graphene. This is the equivalent of a single flat layer of graphite: carbon atoms arranged in hexagonal rings across a vast flat plane. Carbon nanotubes and graphene are set to play an important role in the future of electronics and materials science, and researchers have already made exciting prototype electronic devices using carbon nanotubes, including nanoscale transistors, super-capacity batteries and flexible touch screens.

Left: Geodesic dome, of the type designed by architect Richard Buckminster Fuller. Its structure is similar to the molecule above – but this dome consists only of hexagons, not hexagons and pentagons.

Another allotrope of carbon is amorphous carbon – amorphous (without shape) because the atoms are not arranged in a regular pattern. It is found in charcoal and in soot. Charcoal played an important role in the history of technology: it was used to smelt metals. Activated charcoal is used in water filters and gas masks.

There are more carbon compounds than those of all the other elements combined; more than 10 million have already been studied. Carbon's versatility is due to its ability to form single, double and triple bonds, and rings, and the ease with which it joins with other elements – particularly hydrogen, oxygen and nitrogen. Molecules with just carbon and hydrogen, called hydrocarbons, include methane and propane; candle wax is made of long-chain hydrocarbons. Life as we know it depends upon carbon chemistry, and every living thing ever discovered is a carbon-based life form. Because of the fact that carbon compounds are essential to all living organisms, carbon chemistry is called organic chemistry. However, the importance of carbon compounds extends far beyond the realm of living things. Organic compounds form the basis of the petrochemical industry, and so include plastics and many synthetic dyes, adhesives and solvents.

Carbon compounds are intimately involved in all the processes of life – from photosynthesis and respiration through nutrition and repair to growth and reproduction. In photosynthesis, plants (and some other organisms) use light energy to build the simple sugar glucose ($C_6H_{12}O_6$) from carbon dioxide and water. The result is a store of chemical energy that can be utilized via respiration, by those plants as well as by animals that consume them. Photosynthesizing organisms typically combine the glucose molecules to form larger organic molecules, such as sucrose and starch. In addition to providing the energy that living things need, organic compounds also provide the structural materials from which living things are made. Larger molecules made up from units of glucose include cellulose, used by plants to build cell walls – and one of the main constituents of wood (and therefore paper). Other structural molecules, such as lignin in wood, and structural

Above: Lit wax candle. Candle wax is a mixture of hydrocarbon molecules, with about 30 carbon and 60 hydrogen atoms per molecule. Heat from the flame vaporizes the wax; the carbon and hydrogen combine with oxygen, forming water and carbon dioxide.

Below left: Model of a methane molecule – its tetrahedral shape is due to carbon's sp³ hybrid orbitals (see page 104).

Below: Strand of DNA. The purple and pink strands are the sugar-phosphate "backbone", the yellow, blue, red and green "rungs" are the bases that carry information. All the component parts are organic molecules.

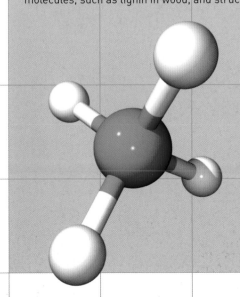

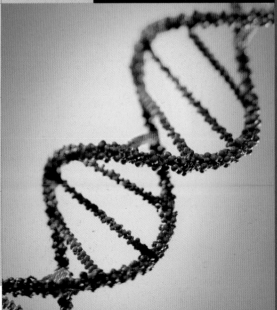

proteins in animals, are also carbon-based. Chemicals involved in maintaining and repairing living organisms include enzymes, hormones and other signalling molecules and anti-oxidants such as vitamins – all are organic molecules. Finally, carbon compounds are at the heart of how living things reproduce, via the large organic molecule DNA (deoxyribonucleic acid).

Carbon is present in all the systems that make up planet Earth – in the biosphere (all living things), the atmosphere, the lithosphere (Earth's crust) and the hydrosphere (rivers, lakes and oceans). The constant, cyclical interchange of carbon between these systems is called the carbon cycle. The main component of this cycle involves carbon dioxide being absorbed from the atmosphere during photosynthesis and then being released back into the atmosphere by respiration. Carbon dioxide is also released from once-living things if they burn. And when an organism dies, it typically decomposes – a process that results in its carbon content being released to the atmosphere as carbon dioxide or methane.

In some circumstances, an organism does not decompose. Its carbon content may slowly form mixtures of hydrocarbons: fossil fuels such as oil and coal. During some epochs of Earth's history, there have been dramatic shifts in the carbon cycle. During most of the last 10 million years, the proportion of carbon dioxide in the atmosphere has remained at around 300 parts per million. However, since humans began burning fossil fuels, building a carbon-based economy, the carbon cycle has been out of balance. In 2012, atmospheric carbon dioxide had reached more than 390 parts per million, and it continues to rise steeply. Since carbon dioxide is a greenhouse gas, increased levels seem to be causing the planet to warm, via an enhanced greenhouse effect. For this reason, scientists and engineers are searching for alternatives to the carbon-based economy (see hydrogen, page 21).

The technique called radiocarbon dating – used to provide good estimates of the age of once-living material – involves the isotope carbon-14. This radioactive isotope is produced at a constant rate in the atmosphere as a result of cosmic ray bombardment. While an organism is alive, its carbon content is constantly replenished, either via photosynthesis or by consuming photosynthetic organisms. The proportion of carbon-14 to the stable isotope carbon-12 is therefore constant when an organism is alive – but once dead, the carbon-14 decays (with a half-life of about 5,700 years). The longer a sample of living material has been dead, the less carbon-14 remains.

Above Oil platform, for extracting oil and natural gas from rocks beneath the ocean. Since the eighteenth century, an estimated 340 billion tonnes of carbon has been released by burning fossil fuels – mostly as part of the compound carbon dioxide.

Right: The world's tropical rainforests store more than 200 billion tonnes of carbon, which the forests' trees and other plants sequestered from atmospheric carbon dioxide, via photosynthesis.

14

Si

Silicon

ATOMIC NUMBER: 14

ATOMIC RADIUS: 110 pm

OXIDATION STATES: **-4**, -3, -2, -1, +1, +2, +3, **+4**

ATOMIC WEIGHT: 28.09

MELTING POINT: 1,410 °C (2,570 °F)

BOILING POINT: 2,355 °C (4,270 °F)

DENSITY: 2.33 g/cm³

ELECTRON CONFIGURATION: [Ne] $3s^2\ 3p^2$

The element silicon is best known for its applications in electronics, as the most widely used semiconductor. Most computer chips ("silicon chips") and other integrated circuits are made from wafers of ultra-pure, silvery crystalline silicon. The silicon is doped with other elements, changing the properties of the silicon to make the transistors and other components on the chip. Ultra-pure silicon is also the basis of around 90 per cent of solar cells, which also make use of silicon's status as a semiconductor.

Silicon is the second most abundant element, by mass, in Earth's crust. It is most commonly found – combined with the most abundant element, oxygen – in minerals called silicates, which contain the silicate ion ($(SiO_4)^{4-}$). Silicate rocks are used to make bricks, ceramics and cement.

   Closely related to the silicate minerals is the compound silica (silicon dioxide, SiO_2), which is found in quartz. When a voltage is

Right: Sample of the metalloid element silicon.

Below: Sand dunes in Algeria. Sand is made of small grains of various minerals, usually predominantly quartz – a silicate mineral.

ordered crystal structures that form if the liquid silica solidifies slowly. Our ancestors used one form of quartz, known as flint, to make axes and arrowheads; the name "silicon" is derived from the Latin word *silex*, meaning "flint". Silicon was first produced in its elemental form – and recognized as an element – in 1824, when Swedish chemist Jöns Jacob Berzelius managed to produce a powder of fairly pure amorphous silicon (like amorphous carbon; see page 107).

The main component of quartz, silica, is also used in a gel that absorbs water from the air; silica gel is often found in small paper packets inside packaging carrying consumer products that must be kept dry. There are many other applications of silicon in the modern world, including its use in organic polymer compounds called silicones, used in rubbery, heat-resistant cookware, and in a liquid or gel form in household sealants and breast implants. Silicon is also a common ingredient in a range of alloys with iron or aluminium; for example, silicon steel is used to make the cores of electrical transformers.

applied to a thin quartz crystal, it bends – and conversely, when the crystal bends, it produces a voltage. This phenomenon, known as piezoelectricity, is put to use in quartz clocks and watches, which contain quartz crystal oscillators that provide a very regularly fluctuating electric signal.

Silica, mostly from quartz sand (quartz being the main component of most sand), is the main ingredient in glassmaking; when molten silica is cooled rapidly, the molecules form disorganized networks, characteristic of glass, rather than the

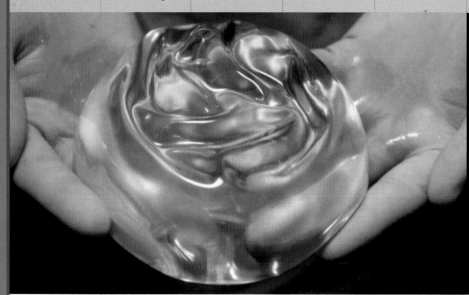

Above: Part of a microprocessor (magnification about x200). Components such as transistors, as well as the (aluminium) connecting tracks between them, are formed by etching into and depositing materials on to ultra-pure silicon.

Left: Breast implant filled with silicone gel. A silicone is a polymer – its large molecules consist of repeated units of small molecules. The starting point of all silicones is silicon dioxide.

32
Ge
Germanium

Like silicon, germanium is a semiconductor – and it, too, is used in electronics, though much less so than silicon. Some integrated circuits ("chips") are made with a mixture of both silicon and germanium, rather than pure silicon, for example. Both elements are metalloids, with some characteristics of metals and some of non-metals.

ATOMIC NUMBER: 32
ATOMIC RADIUS: 125 pm
OXIDATION STATES: -4, +1, **+2**, +3, **+4**
ATOMIC WEIGHT: 72.63
MELTING POINT: 938 °C (1,721 °F)
BOILING POINT: 2,834 °C (5,133 °F)
DENSITY: 5.32 g/cm^3
ELECTRON CONFIGURATION: [Ar] 3d^{10} 4s^2 4p^2

As an element, germanium is hard and whitish-silver in colour. Its existence was predicted by Dmitri Mendeleev, as an element that would fill the gap below silicon in his periodic table; he called it *eka-silicon*. German chemist Clemens Winkler discovered the missing element in a mineral called argyrodite (Ag$_8$GeS$_6$) in 1886. When Winkler analysed the mineral, there was 7 per cent of its mass that he could not account for with the elements then known. In the same year, he managed to produce a fairly pure sample of the new element; he derived the element's name from the Latin name for his home country, *Germania*.

The most important compound of germanium is germanium dioxide (GeO$_2$), called germania. It has various applications, including its use in optical fibres and as an industrial catalyst.

50
Sn
Tin

The element tin was known to ancient metalworkers, and is one of the seven "metals of antiquity" (see page 76); in Ancient Greek and Roman mythology, and in alchemy, tin was associated with the planet Jupiter. The Latin word for tin, *stannum*, is the origin of this element's chemical symbol, Sn.

ATOMIC NUMBER: 50
ATOMIC RADIUS: 145 pm
OXIDATION STATES: -4, **+2**, +4
ATOMIC WEIGHT: 118.71
MELTING POINT: 232 °C (450 °F)
BOILING POINT: 2,590 °C (4,695 °F)
DENSITY: 7.29 g/cm^3
ELECTRON CONFIGURATION: [Kr] 4d^{10} 5s^2 5p^2

The alloy of copper and tin known as bronze has been produced for at least 5,000 years (see page 72). Pewter is another well-known tin alloy, typically consisting of 90 per cent tin with other metallic elements such as antimony, copper, bismuth and lead, which act as hardening agents. Tin is also a component of solder, used

Top: Samples of pure tin.

Above: Sample of the metalloid element germanium.

to connect plumbing pipes and electrical wiring. In its elemental form, tin is applied as a thin protective layer on other metals; most food cans are made from tin-plated steel, for example.

Tin metal can exist in either of two allotropes (see page 105), each with a different crystal structure. Alpha tin (α-tin) is a non-metal and has a dull grey appearance, while beta tin (β-tin) has metallic bonding, with free electrons, behaves like a metal and has a silvery lustre.

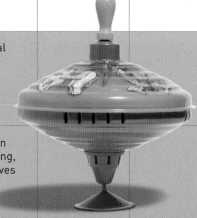

Right: Child's spinning top, made of painted tin. From the 1850s until the rise of plastics in the 1940s, tin was one of the most popular materials for making toys.

82

Pb

Lead

ATOMIC NUMBER:	82
ATOMIC RADIUS:	180 pm
OXIDATION STATES:	-4, **+2**, +4
ATOMIC WEIGHT:	207.20
MELTING POINT:	327 °C (621 °F)
BOILING POINT:	1,750 °C (3,182 °F)
DENSITY:	11.34 g/cm³
ELECTRON CONFIGURATION:	[Xe] 4f¹⁴ 5d¹⁰ 6s² 6p²

ELECTRON CONFIGURATION: [Xe] $4f^{14}$ $5d^{10}$ $6s^2$ $6p^2$

Above right: Ingot of pure lead. Standard cast ingots like this are convenient to sell, and to transport for re-melting and casting into the desired shape.

Like tin, lead was known to ancient metalworkers, and is one of the seven "metals of antiquity" (see page 76); in Ancient Greek and Roman mythology, and in alchemy, lead was associated with the planet Saturn. In its elemental form, lead is a soft, silvery-blue metal with a high lustre.

People were using lead long before the rise of the Roman Empire. Lead is easy to extract from its ores, and there is evidence that small-scale lead smelting began around 9,000 years ago, in some of the earliest settlements in what are now Turkey and Iraq. The Ancient Egyptians also smelted lead – and they used lead compounds in cosmetics, pigments and medicines. A 2010 analysis of Ancient Egyptian eye make-up identified two lead compounds that do not occur naturally; they must have been made on purpose. In laboratory tests, these compounds elicited an immune response; it may well be that the Ancient Egyptians were aware of that fact that their make-up would reduce the chance of eye infections. Several ancient Roman medicines also made use of lead compounds – despite the fact that the Romans were aware of the phenomenon of lead poisoning (see opposite).

Like many metals, lead quickly tarnishes when exposed to the air, giving it a dull grey appearance. The layer of tarnish on metallic lead protects it against corrosion. This, together with the fact that its ductility and malleability make it easy to work with, has made lead suitable as a roofing material for hundreds of years, as it still is today. From the time of the Roman Empire until the second half of the twentieth century, lead was also the material of choice for plumbing pipes. In fact, the word "plumbing" comes from the Latin word for lead, *plumbum* – as does the element's symbol, Pb. The word *plumbum* was a generic term meaning "soft metal"; lead was *plumbum nigrum* and tin was *plumbum album* or *plumbum candidum* (dark and light plumbum); another metal, bismuth, was was sometimes referred to as *plumbum cineareum*. The word "lead" simply comes from an Old English word for the metal.

Lead played a crucial role in the invention of the printing press in the 1430s. The inventor of the press, German goldsmith Johannes Gutenberg, needed a metal that would melt at fairly low temperature, so that it could be cast to make type, but that was hard enough to survive being pressed hard on to paper. Gutenberg started with lead – which was too soft – and ended up with an alloy of lead, tin and antimony.

It is well known that lead is toxic. Symptoms of acute lead poisoning include diarrhoea, kidney damage and muscle weakness. But a greater danger comes from chronic exposure to lead; like many heavy metals, it accumulates in tissues of the body. Over long periods it disrupts the nervous system. It is most dangerous in children, since it delays brain development and gives rise to attention problems. The exact mechanisms of lead's neurotoxicity are still unknown, but recent research has uncovered the fact that lead interferes with the release and the function of proteins called brain-derived neurotrophic factors. These proteins encourage the growth of new neurones, and are vital for learning and memory. As a result of concerns over lead in the environment, once-popular applications of lead have declined or in many cases been banned. Lead-based paints and leaded petrol – with the anti-knocking additive tetraethyl lead $(CH_3CH_2)_4Pb$ – are the two best examples.

Lead is still used as shielding in X-ray machines and as a weight, or ballast, in boats, while lead-acid car batteries account for around half of all lead production. The Romans and the Egyptians obtained their lead from the ore called galena – mainly lead(II) sulfide (PbS) – which is still the main source of the metal today. Around 10 million tonnes of lead are still produced each year worldwide, about half of it from recycled scrap.

Lead exists in three stable isotopes, the most common being lead-208. The nucleus of lead-208 happens to be the heaviest stable nucleus of any element. It is the end product of a decay chain that begins with unstable, radioactive elements such as the actinoid thorium (see page 95); as one unstable nucleus disintegrates, another forms, and the chain only ceases when stable lead-208 is reached. Two other stable lead isotopes, lead-206 and lead-207, are end points of other decay chains. The only lead isotope that has been on Earth since the planet formed is lead-204, which is an unstable isotope – albeit with a half-life of trillions of years.

Above: Wall painting from the tomb of Egyptian Queen Nefertari (thirteenth century BCE). Like most Ancient Egyptian women, Nefertari wore black mascara made from lead sufide (or in some cases, antimony sulfide).

Left: Lead fishing weight and "split" lead shot crimped around fishing lines to weigh them down. Due to lead's toxicity, many anglers now use alternatives to lead for weights.

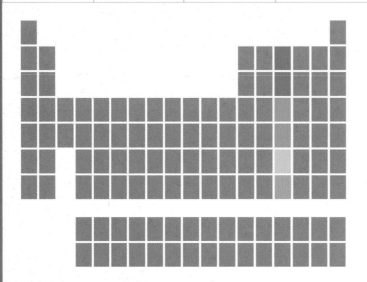

7
N
Nitrogen

15
P
Phosphorus

33
As
Arsenic

51
Sb
Antimony

83
Bi
Bismuth

115
Uup
Ununpentium

The Nitrogen Group

As is true for the elements of Group 14, the elements of Group 15 vary considerably in their properties – in this case, from the non-metals nitrogen and phosphorus, through the metalloids arsenic and antimony, to the "poor metal" bismuth. This group also includes element 115, which has the temporary name "ununpentium" (1-1-5-ium). Element 115 does not occur naturally, but has been created in nuclear laboratories; it features with the other transuranic elements, on pages 153–157.

Nitrogen and phosphorus are extremely important to all life on Earth, and as a result, are both essential ingredients in fertilizers. Nitrogen is a defining ingredient in all proteins and DNA (deoxyribonucleic acid); phosphorus is also a crucial part of DNA, and also of life's system of energy "currency" – the two molecules ATP (adenosine triphosphate) and ADP (adenosine diphosphate). Arsenic and antimony, on the other hand, are highly toxic. Two of these elements – phosphorus and arsenic – exist in at least two forms, or allotropes, when pure (see page 105).

The elements of Group 15 are sometimes referred to as pnictogens. The name derives from the Greek *pnigein*, meaning "choking" or "stifling"; the element nitrogen is inert, and can therefore stifle a flame. The atoms of these elements have five electrons in their outer shell – two in an s-orbital and three in p-orbitals (see pages 8–9);

each element's electron configuration ends with $s^2 p^3$. The s-orbital, containing two electrons, is spherical. The other three electrons occupy one p-orbital each; this particular configuration makes the p-orbitals spherically symmetrical. As a result, when these atoms occur singly, they are spheres. This is rather rare among elements; it is true only for elements with half-filled and filled electron shells. However, these elements rarely exist as single atoms. Even pure nitrogen, a gas at room temperature, exists as two-atom (diatomic) molecules, N_2. The half-filled shells of these elements mean that they readily form bonds with other atoms of their kind or with other elements. The strength of these bonds means that most of their compounds are actually very stable.

7

N

Nitrogen

ATOMIC NUMBER: 7
ATOMIC RADIUS: 65 pm
OXIDATION STATES: **-3**, -2, -1, +1, +2, **+3**, +4, **+5**
ATOMIC WEIGHT: 14.01
MELTING POINT: -210°C (-346°F)
BOILING POINT: -196°C (-320°F)
DENSITY: 1.25 g/L
ELECTRON CONFIGURATION: [He] 2s² 2p³

In its elemental form, at normal temperatures, nitrogen is a colourless, odourless gas. Each nitrogen molecule is composed of two nitrogen atoms; the three half-filled p-orbitals of each atom overlap, forming a highly stable triple bond that makes nitrogen gas very unreactive, or inert. This is why, for example, nitrogen gas is used as an inert atmosphere, often accompanied by carbon dioxide, in situations where the presence of oxygen would be undesirable, such as in food packaging for some fresh salads and meats. Nitrogen is by far the most abundant element in Earth's atmosphere, but only thirty-first most abundant in the crust; this discrepancy is due to nitrogen's inertness, since if it were more reactive, it would combine with other elements in the crust, forming minerals (compare with oxygen; see page 123).

Right: Elemental nitrogen in a gas jar. At normal temperatures and pressures, nitrogen exists as a colourless gas composed of diatomic molecules (artwork inset).

Below: Stick of dynamite – nitroglycerine soaked into clay - with blasting caps, which are inserted into the dynamite so that it can be detonated from a distance.

Despite nitrogen's inertness, its existence was uncovered relatively early in modern chemistry, by scientists studying "airs" (gases). These pioneering chemists had realized that ordinary air is composed of at least two gases – one in which things could burn and animals could breathe, and one in which they couldn't. Around 1770, English scientist Henry Cavendish was studying "burnt air", which would have been composed mostly of carbon dioxide and nitrogen. He removed the carbon dioxide and was left with a gas that he realized constitutes almost 80 per cent of normal air. Cavendish did not publish his results, and over the next few years, several other scientists also isolated the same gas. The first to publish was English chemist, Daniel Rutherford, in 1772.

French chemist Antoine Lavoisier realized that this new "air" must be an element. In 1789, he suggested the name "azote" – the prefix "a-" meaning "without" and "zote" from the Greek word *zoo*, meaning "life" – since no living things can survive in pure nitrogen. Some nitrogen compounds still have "az-" as part of their name to this day. For example, "azo dyes" are synthetic pigments used in a wide range of textiles and other products, while sodium azide (NaN_3) is an explosive compound that rapidly produces large volumes of nitrogen gas to inflate vehicle airbags. As scientists studied the new gas, they realized it could make nitric acid (HNO_3), which in turn could make nitre – the compound potassium nitrate (KNO_3), which was used for making gunpowder. The name *nitrogène* – meaning

GELATIN EXTRA 40%

N°6
BLASTING CAPS
DANGEROUS

Above: Lightning bolts striking the ground. The enormous energy liberated by a lightning bolt drives reactions between nitrogen and oxygen, the two main constituents of the atmosphere. As a result, lightning "fixes" atmospheric nitrogen, forming nitric oxide.

Below: Root nodules on a pea plant (*Pisum sativum*). Each nodule is packed with nitrogen-fixing bacteria.

"nitre-generator" – was suggested by the French chemist Jean Antoine Chaptal in 1790.

From the 1830s, scientists reacted nitric acid with cellulose, the fibrous part of plants. The resulting nitrocellulose found wide application. Nitrocellulose was used to make smokeless explosives (gun cotton) – and, in the 1850s, the first synthetic plastic, Parkesine, invented by English inventor Alexander Parkes. Parkesine developed into celluloid, the material of choice for photographic film for more than 50 years. An offshoot of the production of nitrocellulose was the highly explosive nitroglycerin. In 1867, Swedish chemist Alfred Nobel invented dynamite, by incorporating nitroglycerine into clay, making it safer to transport and easier to use. Nitroglycerine is also used in medicine, to relieve the symptoms of angina – chest pain caused by reduced blood flow to the heart. Taken in tablets, capsules or mouth spray, nitroglycerine breaks down once inside the body to form nitric oxide (NO), which acts to widen constricted arteries, increasing blood flow.

In addition to the demand for nitrogen in explosives and plastics, there was a new need for nitrogen, in fertilizers, after scientists of the nineteenth century realized that it was essential for life. We now know that nitrogen is present in a wide range of molecules essential in all living processes – in particular, DNA and all proteins. Despite nitrogen's abundance in the atmosphere, however, its inertness makes it difficult for living things to get hold of it. In fact, there are not many kinds of organism that can "fix" nitrogen from the air: just a few types of bacteria. Fortunately, nitrogen-fixing bacteria (diazotrophs) are found in abundance in the oceans, rivers, lakes and soil; there are even colonies inside nodules in the roots of certain plants. These organisms break nitrogen molecules' triple bonds, allowing nitrogen to react with hydrogen. The resulting compound, ammonia (NH_3), is nitrogen's main vehicle into the living world. Together, diazotrophs fix an estimated 200 million tonnes of nitrogen worldwide each year. Lightning provides a not insignificant 9 million tonnes per year: it causes nitrogen to react with oxygen, and the resulting nitric oxide (NO) is washed into the soil as nitric acid (HNO_3), which plants can also utilize.

Until the early twentieth century, bacteria and lightning were the only two ways in which living organisms could obtain bioavailable nitrogen. Nitrogen-rich fertilizers and explosives were mostly made from deposits of bird excrement called guano, but this was in short supply, and scientists were concerned that agriculture would not cope with the demand for food for the burgeoning world population.

In 1908, German chemist Fritz Haber developed a process for producing ammonia using atmospheric nitrogen. Haber's main breakthrough was the use of high-pressure reaction vessels. Another German chemist, Karl Bosch, overcame the difficulties of scaling up the process, and ammonia was first produced on an industrial scale in 1913. Today, the Haber-Bosch process fixes around 100 million tonnes of nitrogen each year – most of it used to produce artificial fertilizers. In artificial fertilizers, the ammonia produced by the Haber-Bosch process usually ends up as ammonium nitrate (NH_4NO_3) or urea ($CO(NH_2)_2$). Without artificial fertilizers, the world's farmers would have little hope of supporting Earth's huge human population. Ammonia produced by the process is also used to make plastics, pharmaceuticals and explosives. Some of the ammonia produced by the Haber-Bosch process is used to make nitric acid, which is another compound used in a wide range of important industrial reactions.

Nitrogen itself is clearly not toxic – otherwise every breath we take would be harmful. However, a class of compounds called cyanides are among the most infamous and most rapidly acting poisons. Cyanide molecules include a carbon atom joined to a nitrogen atom with a triple bond; inside the body, cyanide ions ($(CN)^-$) interfere with cellular respiration, which is essential to all cell functions.

Another nitrogen-containing molecule that has fast-acting effects inside the human body is nitrous oxide (N_2O). Commonly known as laughing gas, nitrous oxide is inhaled as an anaesthetic and a pain killer, during labour and in dentistry. Nitrous oxide has many other uses – including as an additive to some fuels, where it provides extra oxygen so that the fuel burns more rapidly. It is also used as a propellant in some aerosols, notably in canisters that produce whipped cream. (Initial experiments to produce spray cream used carbon dioxide, which produced an acidic solution that made the cream curdle.) Another oxide of nitrogen, nitrogen dioxide (NO_2), is a toxic gas. Some is produced from the nitric oxide made by lightning (see above), but it is also produced in the combustion engines of cars and other vehicles. One of the reactions inside a catalytic converter fitted to vehicle exhausts detoxifies nitric oxide by splitting it into nitrogen and oxygen, to reduce this pollution.

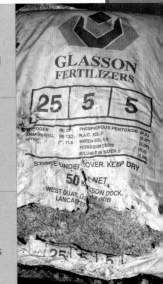

Above: Nitrogen-rich artificial fertilizer spilling out of a torn bag – the three numbers on the bag are the NPK rating, based on the relative proportions of the elements nitrogen (N), phosphorus (P) and potassium (K) present.

Left: Liquid nitrogen being poured into a flask.

15

P

Phosphorus

ATOMIC NUMBER: 15
ATOMIC RADIUS: 100 pm
OXIDATION STATES: -3, -2, -1, +1, +2, **+3**, +4, **+5**
ATOMIC WEIGHT: 30.97
MELTING POINT: 44°C (111°F), white; 610°C (1,130°F), black
BOILING POINT: 281°C (538°F), white. Red phosphorus sublimes.
DENSITY: 1.82 g/cm³, white; 2.30 g/cm³, red; 2.36 g/cm³, black
ELECTRON CONFIGURATION: [Ne] $3s^2\ 3p^3$

The non-metallic element phosphorus was the first element whose discovery is documented, and whose discoverer is known. In 1669, German merchant and part-time alchemist Hennig Brand was trying to extract gold from a yellow liquid: urine. He boiled the urine, to drive off the water, then heated the oily residue to a high temperature. The residue broke down into its various fractions; the last one produced a thick smoke that condensed in water to give a curious white, waxy solid that glowed with an eerie green light. Brand named his new substance *phosphorus mirabilis*, or "miraculous light giver", after the Greek word *phosphoros*, which means "torch-bearer". The glow is the result of a chemical reaction between the phosphorus and oxygen from the air.

Right Samples of red phosphorus, kept in oil. Although not as reactive as white phosphorus, this red allotrope does react with oxygen – violently so at temperatures above about 300°C.

Below: The head of a "strike anywhere" match. These matches contain phosphorus sesquisulfide (P_4S_3). Safety matches do not contain phosphorus or its compounds, but the striking surface does contain red phosphorus, mixed with ground glass.

Phosphorus exists in several allotropes (see page 105). White phosphorus spontaneously ignites at 30°C, and is highly toxic. Since the nineteenth century, it has been used in weapons – for example, in smoke bombs – although its use has been curtailed in recent years through international legislation. White phosphorus was also used in match heads from the early nineteenth century, but concerns over its severe toxicity meant that another allotrope (form) of phosphorus replaced it, in the early twentieth century. Red phosphorus – which is stable, non-toxic and does not produce an eerie glow – is still used in match heads to this day. Two other allotropes of pure phosphorus are known: violet and black.

In nature, phosphorus is nearly always bound to four oxygen atoms, making a phosphate ion ((PO_4)$^{3-}$),which is found in a wide range of phosphate minerals, and in all living things. Phosphate ions also form part of the sugar-phosphate "backbone" of every molecule of DNA (deoxyribonucleic acid). In addition, they are a central component of the living world's "energy currency" system: the interchange between the molecules adenosine diphosphate (ADP) and adenosine triphosphate (ATP). Most of the 750 or so grams of phosphorus in an adult human body is found in bones and teeth, as hydroxylapatite ($Ca_5(PO_4)_3(OH)$), a form of the mineral calcium phosphate.

Most fertilizers and many animal feeds contain significant amounts of phosphates; fertilizers are given an NPK rating, based on the proportions of the elements nitrogen (N), phosphorus (P) and potassium (K) present. The extraordinary demand for mineral phosphates to help feed the world's huge population may put resources under pressure, and some mining experts have warned that supplies may dwindle in a matter of decades without a huge increase in the recycling of this important mineral – in particular, from sewage.

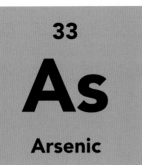

33

As

Arsenic

ATOMIC NUMBER: 33

ATOMIC RADIUS: 115 pm

OXIDATION STATES: -3, +2, **+3, +5**

ATOMIC WEIGHT: 74.92

MELTING/BOILING POINT: Sublimes at 615ºC (1,137ºF)

DENSITY: 5.73 g/cm³

ELECTRON CONFIGURATION: [Ar] $3d^{10} 4s^2 4p^3$

Right: Elemental arsenic, a toxic metalloid.

Below: Arsenic is one of a small number of elements that occurs naturally, or natively, in its elemental state. This specimen is the so-called botryoidal (grape-like) form of native arsenic.

Pure arsenic occurs as several allotropes. The contrast between the two most important ones highlights arsenic's metalloid qualities: grey arsenic is dense and lustrous like metals, while yellow arsenic is a crumbly powder like many non-metals. Arsenic sublimes (turns directly from a solid to a gas) at 614°C at standard atmospheric pressure. Under much higher pressure, arsenic can exist as a liquid; its melting point at such high pressure is 817°C – higher than its boiling point under normal circumstances.

Arsenic is sometimes found in its native (uncombined) state, but is much more commonly found in a range of minerals. Arsenic minerals were used as pigments from ancient times right up to the end of the nineteenth century. The ancients' most widely used arsenic mineral pigment was orpiment, composed chiefly of golden yellow arsenic trisulfide (As_2S_3). The ancient Persian word for orpiment was *zarnik*, which means "gold-coloured", and that word is the ultimate root of the word "arsenic". In addition to using arsenic compounds in pigments, many ancient civilizations used them in medicines. Arsenic medicines were popular in the nineteenth century, and even today some arsenic compounds are used as anti-cancer drugs.

The earliest bronzes (see copper, page 72), dating to around 4000 BCE, were actually alloys of copper and arsenic rather than tin. Several ancient and medieval scholars and alchemists reported obtaining a metallic substance from arsenic minerals, but arsenic was not really considered an element until the advent of modern chemistry. Today, the main use of metallic arsenic is for alloying with copper and lead. It is also used in electronics, to make the compound gallium(III) arsenide (GaAs), which is used to make some integrated circuits and light-emitting diodes (see gallium, page 102).

Pure arsenic and most arsenic compounds are highly toxic, and were the poison of choice for murder and suicide from medieval times until the beginning of the twentieth century, when better detection and antidotes became available. Two of the biggest uses of arsenic compounds in the twentieth

century were in wood preservatives and insecticides. But arsenic's toxicity has caused these applications to be banned in many countries. During the 1990s, international aid agencies and the World Health Organization became concerned about the use of groundwater contaminated with arsenic. It followed the installation of millions of wells dug tens of metres deep, with the intention of providing an alternative to water from above ground, which was often contaminated with disease-causing bacteria and other pathogens. However, in many places, the rocks beneath the ground leached arsenic compounds into the water and caused hundreds of thousands of cases of arsenic poisoning. Estimates vary, but more than 100 million people in the world today are exposed to dangerous levels of arsenic in this way.

Above: A geothermal pool in New Zealand, fed by hot water rising up as a result of volcanic activity. Water in geothermal pools is rich in dissolved and precipitated minerals. This pool is rich in arsenic sulfide, as well as sulfides of antimony and mercury.

The Group 15 element antimony is very similar to its cousin arsenic: it is a metalloid, it occurs in several different allotropes, and it is highly toxic and was long used as both a poison and a medicine. Also like arsenic, antimony is sometimes found native (uncombined), but exists more commonly in minerals.

51

Sb

Antimony

ATOMIC NUMBER:	51
ATOMIC RADIUS:	145 pm
OXIDATION STATES:	-3, +3, +5
ATOMIC WEIGHT:	121.76
MELTING POINT:	631°C (1,168°F)
BOILING POINT:	1,587°C (2,889°F)
DENSITY:	6.69 g/cm³
ELECTRON CONFIGURATION:	[Kr] 4d¹⁰ 5s² 5p³

Right: Elemental antimony. Most antimony is produced from the ore stibnite (antimony sulfide, Sb_2S_3), by heating it with scrap iron. The iron combines with the sulfur, leaving behind elemental antimony.

Antimony and its compounds have been known for thousands of years. In Ancient Egypt, a black mineral containing antimony sulfide (Sb_2S_3) was used as mascara, and the Egyptian word for this mineral, *sdm*, is the ultimate source of the element's name and its symbol. Through the ages, the word was borrowed and changed: in Greek, it became *stimmi*, and in turn, this became Latinized to *antimonium*, which gives the element its name; an alternative Latin term, *stibium*, gives the element its symbol.

Today, metallic antimony is usually obtained as a by-product of copper smelting. Production fluctuates with the world economic situation, but is currently around 180,000 tonnes; China is by far the biggest producer. Antimony is used to make alloys, most often with lead – for example, in the electrodes of car batteries. The most important compound of antimony, antimony oxide (Sb_2O_3), is added to the plastic PVC (polyvinyl chloride) as a flame retardant.

83

Bi

Bismuth

ATOMIC NUMBER: 83
ATOMIC RADIUS: 160 pm
OXIDATION STATES: -3, **+3**, +5
ATOMIC WEIGHT: 208.98
MELTING POINT: 271°C (521°F)
BOILING POINT: 1,564°C (2,847°F)
DENSITY: 9.72 g/cm³
ELECTRON CONFIGURATION: [Xe] $4f^{14}$ $5d^{10}$ $6s^2$ $6p^3$

In its elemental form, bismuth is a dense and shiny silver-white substance, like a metal – although it is more brittle than true metals and it conducts heat and electricity very poorly. (It is sometimes referred to as a "poor metal".) A fresh sample of pure bismuth slowly forms a very thin layer of bismuth oxide when left in the air; this layer gives the metal a pinkish or even multi-coloured iridescent sheen. Until recently, bismuth was thought to have just one stable isotope – bismuth-209. But in 2003, physicists in France discovered that even bismuth-209 is unstable, and decays with an extraordinarily long half-life of 20 million million million years. So in fact, bismuth has no stable isotopes at all.

Right: Samples of the element bismuth, a "poor metal".

Bismuth has few major applications. It is used in various alloys, and its compounds are used in some medicines – the most important being bismuth subsalicylate ($C_7H_5BiO_4$), which is prescribed in liquid or tablet form to treat diarrhoea and heartburn – and cosmetics. It is slightly toxic, but nowhere near as much as arsenic and antimony. The origin of the name "bismuth" is uncertain; it probably comes from a German term meaning "white matter". Alternatively, it is possible that the name originates in the Greek word *psimythion*, meaning "white lead".

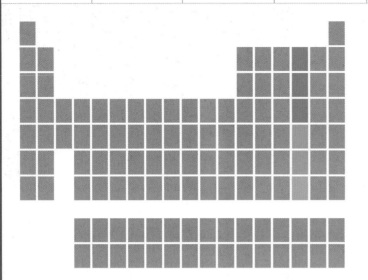

| 8 |
| O |
| Oxygen |

| 16 |
| S |
| Sulfur |

| 34 |
| Se |
| Selenium |

| 52 |
| Te |
| Tellurium |

| 84 ☢ |
| Po |
| Polonium |

| 116 ☢ |
| Lv |
| Livermorium |

The Oxygen Group

The most important and abundant elements of Group 16 of the periodic table are the non-metals oxygen and sulfur. Selenium is also a non-metal, while tellurium is a metalloid. Polonium is sometimes considered a true metal, sometimes a metalloid. It is a radioactive element, with no stable isotopes. This group also includes livermorium, created in nuclear laboratories – see the transuranium elements, pages 153–7.

The elements of Group 16 are sometimes referred to as "chalcogens". The name derives from the Greek chalkos, meaning "bronze" or just "metal"; the elements oxygen and sulfur are present in nearly all metal ores. The atoms of these elements have six electrons in their outer shell, comprising two in an s-orbital and four in p-orbitals; each element's electron configuration ends with $s^2 p^4$. Normally, having the same outer electron configuration would mean that elements have similar properties – that was the basis on which the periodic table was originally conceived. The properties of Group 16, however, are perhaps the most varied of any group, and that is partly due to the electron configuration. To attain a full outer electron shell, an element in this group can either gain two electrons, lose four (to leave just the two s-electrons) or lose all six. Exactly what the atoms do depends upon the particular circumstances, and also on the size of the atom. Oxygen, at the top of the group, has a small atomic radius; as a result, the outer electrons are relatively close to the nucleus and are not easily lost. Instead, oxygen atoms easily gain electrons, giving this element the properties of a non-metal. Much larger polonium (at the bottom) loses electrons fairly easily; this is why polonium is a metallic element. Oxygen is by far the most important element in this group. It is the most abundant element in Earth's crust and in the hydrosphere (Earth's water) and second most abundant in the atmosphere. Oxygen atoms also account for a large proportion of the mass of all living things – 65 per cent in the case of the human body.

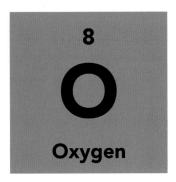

8

O

Oxygen

ATOMIC NUMBER: 8

ATOMIC RADIUS: 60 pm

OXIDATION STATES: **-2**, -1, +1, +2

ATOMIC WEIGHT: 16.00

MELTING POINT: -219°C (-362°F)

BOILING POINT: -183°C (-297°F)

DENSITY: 1.43 g/L

ELECTRON CONFIGURATION: [He] $2s^2 2p^4$

Element number 8, oxygen, is the third most abundant element in the Universe, after hydrogen and helium. Oxygen nuclei are produced deep inside most stars; the most common isotope, oxygen-16, is formed by the fusion of four helium-4 nuclei. Oxygen is the second most abundant element in Earth as a whole, after iron – and by far the most abundant element in Earth's crust, where it occurs mostly in oxide, silicate (SiO_2), carbonate (CO_3) and sulfate (SO_4) minerals.

OXYGEN

Oxygen also accounts for just under 90 per cent of the mass of pure water, and dissolved oxygen increases still further the amount of the element present in water. In addition, oxygen is the second most abundant element in the atmosphere, after nitrogen. There are about one quadrillion tonnes (10^{15}t) of elemental oxygen in Earth's atmosphere, most of it in the form of diatomic molecules (O_2).

Oxygen accounts for 21 per cent of the volume of dry air; the constitution of the atmosphere is normally given for dry air because the amount of water vapour in air varies considerably.

Oxygen is one of the most reactive of all the elements; that is why so much of it exists combined with other elements – in water, rocks and carbon dioxide, for example. The only reason the atmosphere and oceans contain large amounts of elemental oxygen is that certain living things constantly replenish it, as a waste product of photosynthesis. Plants and certain types of bacteria photosynthesize, using the energy from sunlight (*photo* is Greek for "light") to build ("synthesize") the molecules they need to survive. Oxygen is produced in the first stage of the process, when light energy splits water molecules into hydrogen ions (H^+), oxygen gas and free electrons. The hydrogen ions and the free electrons take part in further reactions with carbon dioxide, in which vital, energy-storing organic molecules are synthesized – beginning with carbohydrates such as glucose ($C_6H_{12}O_6$). Oxygen atoms are also present in many other molecules involved in living processes, including all proteins, fats and DNA (deoxyribonucleic acid).

The first photosynthetic organisms evolved on Earth more than three billion years ago. For the first few hundred million years, most of the oxygen that they produced reacted with iron dissolved in the oceans. The result was insoluble iron oxide, which is visible today in rust-coloured banded iron formations found in ancient sedimentary rocks.

Above: Elemental oxygen in a gas jar. At normal temperatures and pressures, oxygen exists as a colourless gas composed of diatomic molecules (artwork inset).

Left: Dense rainforest in Costa Rica. Equatorial rainforests produce around 100 tonnes of oxygen per hectare per year, using energy from sunlight to split water - the first stage of photosynthesis.

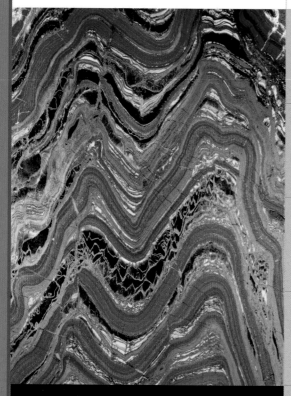

Above: Banded iron formation. The red layers are iron oxide produced when oxygen combined with dissolved iron in oceans around 2.5 billion years ago. Iron oxide is insoluble, so it fell to the ocean bottom as sediment.

Below: Artist's impression of a scene from the Carboniferous period. Extremely large insects thrived in the oxygen-rich atmosphere.

Gradually, however, the concentration of oxygen in the air did begin to increase. The oxygenation of the atmosphere may also have helped cause a 300-million-year-long ice age, which began around 2.4 billion years ago. According to this theory, the free oxygen reacted with atmospheric methane (CH_4), producing carbon dioxide. Since carbon dioxide is a less potent greenhouse gas than methane, Earth's temperature decreased.

The concentration of oxygen in the atmosphere increased dramatically when plants began colonizing the land, around 450 million years ago – and not long afterwards, animals evolved to live on land in the oxygen-rich air. Around 300 million years ago, during the Carboniferous period, the atmospheric concentration of oxygen reached an all-time high of about 35 per cent, due to the fact that nearly all the world's land areas were covered in dense forests of towering trees, all producing oxygen. During the Carboniferous period, many species of insects grew much larger than any insects do today, almost certainly as a result of the oxygen concentration. The size of insects is limited by the way they breathe: through openings called spiracles that connect to tubes inside their bodies. Oxygen diffuses from these tubes directly into the insects' cells, so the larger the insect, the bigger the tubes must be. For an insect with a length of about 30 centimetres in today's atmosphere, the tubes would have to be so large that there would be more tube than insect. But in the oxygen-rich atmosphere of the Carboniferous, insects and spiders could be bigger, because the tubes could be smaller – since the rate of oxygen diffusion into cells depends upon the concentration of the gas in the air. So, in the Carboniferous period, some flying insects had wingspans of nearly a metre and cockroaches 10 times the size of those that exist today. There were also spiders with a "legspan" of around 50 centimetres.

Another result of the build-up of oxygen in the atmosphere was the production of the other common form (allotrope) of elemental oxygen: ozone (O_3). A tiny fraction of the total atmosphere, most ozone is found between 20 and 30 kilometres above ground – the ozone layer – where it intercepts most of the ultraviolet radiation that would be harmful to living things on Earth. When a photon of ultraviolet radiation strikes an ozone molecule (O_3), the molecule splits into oxygen (O_2) and a highly reactive oxygen atom (O). Almost immediately, the oxygen atom recombines with an oxygen molecule, to re-form a molecule of ozone, a process that produces heat. The ultimate result of the process is the conversion of the potentially harmful ultraviolet photon to heat. Ozone is actually toxic, and at lower levels it is considered a pollutant. It is a major component of the photochemical smog that often lies over large cities on sunny days: this is the result of sunlight catalysing (speeding up) the breakdown of nitrogen oxides.

Oxygen is an essential part of combustion (burning). Most of the things we commonly burn – including wood, candle wax and fossil fuels – are rich in carbon and hydrogen. When they burn, the hydrogen (H) combines with oxygen from the air to produce water (H_2O), and the carbon (C) combines with oxygen from the air to produce carbon dioxide (CO_2). Both of these reactions are exothermic (they release energy), which is why burning produces heat. Inside the cells of most living things, a series of controlled combustion reactions occurs, called aerobic respiration, by which the organisms gain the energy they need to survive. Aerobic respiration is the reverse of photosynthesis: in photosynthesis, oxygen and carbohydrates are created from water and carbon dioxide (see page 123, above), while in respiration oxygen reacts with carbohydrates inside cells to produce carbon dioxide and water. Even plants respire, to gain access to the energy they have stored in carbohydrates via photosynthesis.

It was oxygen's role in respiration and combustion that led to the element's discovery, in the 1770s. English chemist Joseph Priestley was the first to publish anything about the discovery, in 1775. A year earlier, he had produced the gas, by heating mercury calx (mercury(II) oxide, HgO), and had discovered some of its properties. Priestley discovered that mice would survive for much longer – and that things would burn much more quickly – in the new "air" than in ordinary air. He also realized that plants produce the same gas. Swedish chemist Carl Wilhelm Scheele had carried out the same experiments as Priestley as early as 1772, but his research was not published until 1777. It was French chemist Antoine Lavoisier who realized that oxygen is a chemical element, and who gave the element its name. In 1774, he too produced oxygen from mercury calx. Three years later, he did the reverse – heating mercury in air to produce mercury calx. Lavoisier noted that some of the air had been used up, and that the mercury calx weighed more than the original mercury. Later, he decomposed water into hydrogen and oxygen. These experiments laid the foundation of modern chemistry, helping Lavoisier to discover that mass is conserved in chemical reactions. Lavoisier mistakenly believed that all acids contained the new element, so he came up with the name *oxygène*, meaning "acid former".

Being so abundant and reactive, oxygen is involved in some way in most processes, both natural and industrial – albeit mostly in

Top left: Satellite image of the "ozone hole" (purple) – a thinning of the ozone layer above Antarctica. Chlorofluorocarbons (CFCs) released from aerosols and refrigerators caused much of the ozone depletion.

Above: Aerial view of photochemical smog over Los Angeles, California, USA. A major component of this kind of smog is low-level ozone, which irritates and even damages the human respiratory system.

Below: One of the modules at the Halley Research Station in Antarctica. It was at this research station that the ozone hole was discovered, in 1985. Scientists at the station still monitor the state of the ozone layer to this day.

compounds. So, for example, many metal ores are oxides of the metal, in which case extraction of the pure metal involves removing the oxygen. Iron production (see page 59) begins with iron oxide (Fe_2O_3 or Fe_3O_4), which is heated to a high temperature inside a blast furnace together with a source of carbon (normally coke). The iron oxide gives up the oxygen, which combines with the carbon to produce carbon dioxide. The resulting iron normally contains so much carbon that it is brittle. To burn off the carbon, as well as impurities such as sulfur, an oxygen lance blows pure oxygen into a conversion vessel.

Millions of tonnes of oxygen are produced industrially each year. More than half of it is used in steelmaking, and much of the rest in the chemical industry. Oxygen is also used in the oxy-fuel welding of metals; typically, a supply of pure oxygen is combined with a fuel to produce a very intense high-temperature flame. The most common fuel is ethyne (C_2H_2), also known as acetylene. A similar process is sometimes used to cut metals: a flame heats the metal to high temperature, so that when oxygen is directed on to the surface, the metal burns, producing a liquid metal oxide waste that is blown away, leaving a clean cut in the metal.

Some space-bound rockets carry liquid oxygen, held separate from the fuel (propellant). Since oxygen is required for combustion, and the rate of burning is limited by the supply of oxygen, carrying pure oxygen allows the fuel to burn extremely rapidly – and to burn when the rocket has risen above the atmosphere. Some rockets derive the oxygen they need from a compound called an oxidizer, which releases large quantities of the gas when it heats up. The oxidizer is normally mixed in with the propellant, but in some cases, the propellant and the oxidizer are chemically combined. Explosives contain an oxidizer, too; in traditional gunpowder, for example, the oxidizer is saltpetre (potassium nitrate, KNO_3).

Other uses for elemental oxygen include supplemental oxygen for patients in hospitals and, mixed with other gases, as breathing gas for people working at high altitude and underwater. The breathing gas supplied to passengers in emergency situations on board commercial airliners is produced chemically on demand, from oxygen-rich compounds such as sodium chlorate ($NaClO_3$) and barium peroxide (BaO_2).

Pure oxygen is produced on an industrial scale mostly by fractional distillation of air. In this process, liquefied air is allowed to warm slowly, and each constituent of the air boils off at a different temperature; liquid oxygen boils at −183°C. Nitrogen and oxygen account for around 99 per cent of the air; the nitrogen boils off first, leaving the nearly pure oxygen behind. Liquid oxygen may be kept cold and liquid in large vacuum flasks, or evaporated to form a gas, which is then compressed and distributed in pressurized containers. Although oxygen gas appears colourless, it is actually a very pale blue (although this is not why the sky is blue) and it is weakly magnetic. Liquid oxygen occupies less than one-eight-hundredth the volume that the gas does, and is a little denser than water. Like the gas, it has a noticeable blue tinge, and it is attracted to the poles of very strong magnets. Liquid oxygen freezes at −219°C, to form solid oxygen, which is also pale blue and has a density about the same as polycarbonate (from which DVDs are made).

Above: Cutting torch being used to cut through a sheet of metal. A high-temperature flame is produced by mixing oxygen with fuel, and another burst of oxygen causes the metal to burn away, forming the cut.

Right: Liquid oxygen at the bottom of a double-walled Dewar flask. Liquid and solid oxygen have a distinct blueish tinge.

16

S

Sulfur

ATOMIC NUMBER: 16

ATOMIC RADIUS: 100 pm

OXIDATION STATES: **-2**, -1, +1, **+2**, +3, **+4**, +5, **+6**

ATOMIC WEIGHT: 32.07

MELTING POINT: 115°C (239°F)

BOILING POINT: 718°C (445°F)

DENSITY: 2.00 g/cm³ (slight variations depending on allotrope)

ELECTRON CONFIGURATION: [Ne] 3s² 3p⁴

Sulfur is one of the few elements found native (in its elemental state), normally near volcanoes or hot springs. Altogether, pure or combined, sulfur is the seventeenth most abundant element in Earth's crust and the tenth most abundant in the Universe as a whole. Sulfur is an essential element for all living things; in particular, it is present in many proteins. The human body contains around 140 grams of it. (There is no recommended dietary intake for sulfur; simply eating enough protein will suffice.)

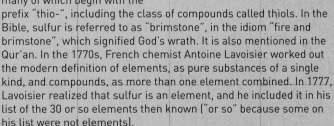

There are more allotropes (different forms) of the element sulfur than of any other element. The most common of the 30 or so known ones are two bright yellow crystalline solids composed of molecules each consisting of eight sulfur atoms. Heated to around 200°C, both these solids melt to form a deep red viscous liquid, in which the eight-atom molecules remain intact.

The name of the element is simply the Latin word for it, *sulpur* or *sulfur*. The Ancient Greeks called sulfur *theion*, a word that persists in several sulfur-containing compounds, many of which begin with the prefix "thio-", including the class of compounds called thiols. In the Bible, sulfur is referred to as "brimstone", in the idiom "fire and brimstone", which signified God's wrath. It is also mentioned in the Qur'an. In the 1770s, French chemist Antoine Lavoisier worked out the modern definition of elements, as pure substances of a single kind, and compounds, as more than one element combined. In 1777, Lavoisier realized that sulfur is an element, and he included it in his list of the 30 or so elements then known ("or so" because some on his list were not elements).

For hundreds of years, the English name of the element has been spelt in two ways – "sulfur" and "sulphur". In America, "sulfur" prevailed by the late nineteenth century, while in Britain, "sulphur" was the norm until 1990, when the International Union of Pure and Applied Chemistry decreed that all English-speaking chemists should use "sulfur".

Because it is found native, sulfur was known to people in ancient civilizations. Even before recorded history began, people used bright orange-red mercury(II) sulfide (HgS), also known as cinnabar, as a pigment to paint on cave walls. Cinnabar also fascinated alchemists,

Above: Lumps of sulfur being heated and forming a thick, deep red liquid.

Above right: Natural (native) deposits of crystalline sulfur in a volcanic area. The crystals are formed when hydrogen sulfide gas from the volcanic vents condenses in the cooler air.

procedure, now known as vulcanization, was rediscovered in 1839 by American inventor Charles Goodyear. Rubber is a natural polymer – composed of long molecules, each made up of repeating units of smaller molecules. Sulfur atoms form bonds between the long polymer molecules, called crosslinks, which are responsible for the improved elasticity and stability. Several synthetic polymers are routinely vulcanized to make them suitable for inclusion in consumer products, from vehicle tyres to silicone rubber switches and the soles of shoes.

Sulfur coated the heads of the earliest matches, made in China in the sixth century. It has remained a significant ingredient in match heads for most of history since then, and a small amount is still included in most modern safety matches. In the ninth century, monks and alchemists in China stumbled upon another incendiary use for sulfur: as an ingredient of gunpowder. Traditional "black powder" is composed of a source of carbon (normally powdered charcoal), saltpetre (potassium nitrate, KNO_3) and sulfur. In the mixture, the sulfur reduces the temperature needed for combustion, enables the powder to burn more rapidly and, since sulfur itself burns, adds to the overall combustibility.

because heating it brings forth beads of shiny mercury metal; it is one of the main ores of mercury to this day (see page 81). Sulfur was sometimes burned in religious ceremonies in several ancient civilizations, to ward off evil spirits, and burning sulfur was also used to fumigate houses, ridding them of insects and other infestations. Today, gardeners still use pure "dusting sulfur" as a pesticide.

In Central America, some people heated sulfur with natural rubber, which made the rubber less sticky and more elastic and durable. This

Sulfur has long been mined from the immediate vicinity of active volcanoes and hot springs. Today, this dangerous practice still goes on in a few places, notably in Indonesia, where the workers' health is in serious danger from poisonous gases such as sulfur dioxide, as well as the intense heat. But most of the world's sulfur production involves removing the element from fossil fuels, where it is an unwanted impurity. Fuels with high sulfur content produce sulfur dioxide – a major pollutant and the

Top left: Red onion slices. Cutting onions breaks cell walls, initiating chemical reactions that result in a sulfur compound called *syn*-propanethial-S-oxide (C_3H_6OS). It is this volatile compound that irritates eyes; tears are released in an attempt to flush it away.

Far left: Trees killed by acid rain. A major cause of acid rain is sulfur dioxide released by coal-fired power stations, which dissolves in cloud droplets, forming an acidic solution.

Left: Men carrying baskets of native sulfur removed from deposits around a lake in the crater of Kawah Ijen volcano, Java, Indonesia.

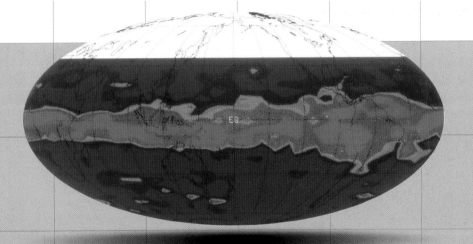

cause of most acid rain – when they burn. Sulfur is also extracted as a by-product of the smelting of some metal ores. One of those ores is pyrite, a mineral of iron sulfide (FeS_2). Throughout history, some miners have mistaken pyrite for gold, due to its metallic lustre and the fact that it sometimes has a golden hue – earning it the nickname "fools' gold". Chinese alchemists discovered that they could extract sulfur from pyrite as long ago as the third century.

There are several very important compounds of sulfur. Sodium metabisulfite ($Na_2S_2O_5$) is widely used as a disinfectant and preservative in winemaking and beer brewing. This convenient powder releases sulfur dioxide when it dissolves in water. Sulfur dioxide is normally a gas, and in that form it is used to create atmospheres in which some soft fruits are stored and transported, to keep them fresh for longer.

The main use for sulfur dioxide – and, in fact, for sulfur itself – is as the precursor to one of the most important compounds in the modern chemical industry: sulfuric acid. World production of sulfuric acid amounts to more than 200 million tonnes each year. The acid has many uses, including in the manufacture of fertilizers, detergents, dyes and some medicines, and as the electrolyte in lead-acid car batteries. Another important sulfur compound is sulfur hexafluoride (SF_6), which is a gas at ordinary temperatures. It is used as an insulator in huge switches at power stations and electricity substations, and is also used as the gas filling some double-glazed windows.

Although pure sulfur is odourless and has very low toxicity, many sulfur-containing compounds are notoriously smelly and some are toxic. The smell of onions and garlic – as well as the unpleasant odour of flatulence – are all due to organic (carbon-containing) sulfur compounds. Butanethiol (also known as butyl mercaptan, $C_4H_{10}S$) is added to natural gas, which has no odour itself, to give people early warning of gas leaks. Other thiols, including ethanethiol (ethyl mercaptan, C_2H_6S), are components of the foul-smelling liquid secreted by the glands under a skunk's tail. The gas sulfur dioxide (SO_2) is noxious and poisonous in high concentrations. Hydrogen sulfide (H_2S) is the familiar smell of rotten eggs, and is also poisonous in high concentrations.

Top: Volcanoes release huge amounts of sulfur dioxide into the atmosphere, as indicated by the red and green areas in this NASA satellite image taken after the eruption of Mount Pinatubo, Philippines.

Right: Wine bottles in a bottling machine. Sodium and potassium metabisulfites are commonly used to sanitize equipment. In the wine itself, sulfur dioxide released by these compounds kills bacteria and acts as an antioxidant, preserving flavours and colours.

34

Se

Selenium

ATOMIC NUMBER: 34

ATOMIC RADIUS: 115 pm

OXIDATION STATES: -2, +2, +4, +6

ATOMIC WEIGHT: 78.96

MELTING POINT: 221ºC (430ºF)

BOILING POINT: 685ºC (1,265ºF)

DENSITY: 4.50 g/cm³ (slight variations depending on allotrope)

ELECTRON CONFIGURATION: [Ar] $3d^{10}$ $4s^2$ $4p^4$

The element selenium is very occasionally found in its native (uncombined) state. It normally occurs in small amounts in ores containing sulfur, including pyrite (iron sulfide, FeS_2). It was from a sample of pyrite that selenium was discovered, in 1817. Swedish chemists Jöns Jacob Berzelius and Johann Gahn were part owners of a sulfuric acid factory in Mariefred, Sweden, and were studying the residue left behind when sulfur from pyrite was burned in the factory's furnace. The residue contained a red substance that they mistook at first for the element tellurium. After further analysis, Berzelius realized the mistake, and announced the discovery of a new element. He named it after the Greek goddess of the Moon, *Selene*, as tellurium, discovered 30 years earlier, had been named after the Roman god of the Earth. Spanish polymath Arnaldus de Villa Nova had actually discovered the same red residue of sulfur around 500 years before Berzelius, but had not, of course, realized that it was a chemical element.

Right: Sample of the grey metalloid form of selenium. The other main allotrope is a red non-metal.

Below: Dietary supplement pills containing several vitamins and selenium. The element selenium has several roles in the body, including being part of a group of enzymes that prevent damage inside cells.

Elemental selenium exists in two common forms (allotropes): a red non-metal (the substance Berzelius and Gahn found) and a dull grey, semiconducting metalloid. When light falls on grey selenium, it changes the element's electrical resistance, because the light photons excite electrons to the conduction band – a set of higher energy levels in which electrons are effectively free of their atoms, as in a metal. The discovery of this phenomenon led to selenium being used in applications such as photographic light meters, photocopiers, fax machines and CCDs (charge-coupled devices, used to capture images in digital cameras). Selenium is also used in digital X-ray machines – it produces a voltage when X-ray photons strike it, a signal that can be used to construct an image.

The demand for selenium for industry is small and generally diminishing. In photocells, it has been largely superseded by other semiconductors, for example – although a thin film of copper indium gallium selenide (CIGS) is used in an increasingly popular type of solar cell. Less than 2,000 tonnes of the element are produced annually, most of it as a by-product of copper extraction. Its applications in glassmaking and the extraction of manganese each account for about one-third of the supply of selenium. In glassmaking, the red pigment selenium dioxide (SeO_2) is used to correct the yellow-green coloration that normally occurs in glass due to the presence of iron impurities; used in larger quantities, it is used to produce red-coloured glass. In architectural glass, it reduces the amount of ultraviolet radiation transmitted. Elemental manganese can be leached out of a solution of its ore with selenium dioxide (SeO_2) added – although selenium is toxic in high doses, and wastes from this process must be carefully controlled.

Despite its toxicity, selenium is an essential element in animals and some plants. In animals, it is involved in the action of several enzymes, and also in the action of some antioxidant vitamins. Foods rich in selenium include Brazil nuts, tuna, turkey and sunflower seeds – albeit only if the plants or turkeys are raised in a selenium-rich environment.

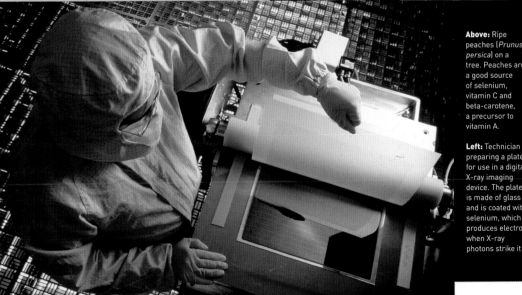

Above: Ripe peaches (*Prunus persica*) on a tree. Peaches are a good source of selenium, vitamin C and beta-carotene, a precursor to vitamin A.

Left: Technician preparing a plate for use in a digital X-ray imaging device. The plate is made of glass and is coated with selenium, which produces electrons when X-ray photons strike it

52

Te

Tellurium

ATOMIC NUMBER: 52

ATOMIC RADIUS: 140 pm

OXIDATION STATES: -2, +2, +4, +5, +6

ATOMIC WEIGHT: 127.60

MELTING POINT: 450°C (841°F)

BOILING POINT: 988°C (1,810°F)

DENSITY: 6.24 g/cm³

ELECTRON CONFIGURATION: [Kr] $4d^{10}\,5s^2\,5p^4$

When pure, the element tellurium has a silvery lustre that makes it look like a metal. In fact, it is a semiconducting metalloid, and, as with its partner selenium (see page 130), its electrical conductivity changes when light falls on it. Thin films of cadmium telluride (CdTe) are used in some solar cells. The main use of tellurium, however, is in alloys with copper and lead, and the element is sometimes added to steel to make it more workable. Tellurium and some of its compounds are extremely toxic.

Right: Native (naturally uncombined) tellurium embedded in quartz rock.

Below: Sample of tellurium. When pure, this element is a metalloid and has a silvery lustre like a true metal.

Tellurium was discovered in 1782 by the Austrian mineralogist Franz-Joseph Müller von Reichenstein, in a sample of gold ore. Von Reichenstein realized another metal was present, in addition to the gold. Several scientists believed the metal contained in the ore to be antimony, but its exact nature was so perplexing that the ore gained the nickname *aurum problematicum*. Von Reichenstein was convinced that it contained a new element, and he was proved right in 1798, when German chemist Martin Klaproth isolated pure tellurium from the ore. Klaproth named the new element after the Latin word *tellus*, meaning "Earth", the following year. Tellurium is extremely rare in Earth's crust – about as abundant as gold. Only a few tens of tonnes of elemental tellurium are produced each year, mostly as a by-product of copper extraction.

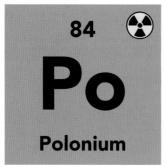

84

Po

Polonium

ATOMIC NUMBER: 84
ATOMIC RADIUS: 190 pm
OXIDATION STATES: -2, +2, +4, +6
ATOMIC WEIGHT: (210)
MELTING POINT: 254ºC (489ºF)
BOILING POINT: 962ºC (1,764ºF)
DENSITY: 9.20 g/cm³ (slight variations
depending on allotrope)
ELECTRON CONFIGURATION: [Xe] 4f¹⁴ 5d¹⁰ 6s²
6p⁴

Right: Sample of the uranium ore pitchblende. Uranium ores are the only places where polonium exists naturally, in minuscule amounts, as the result of the decay of unstable nuclei present in the ore.

Below: Discs containing the radioactive isotope polonium-210, a powerful emitter of alpha particles.

Below right: In November 2006, Polonium-210 was named as the poison used to murder the former KGB agent Alexander Litvinenko in London.

The highly radioactive element polonium was discovered in 1898, by the Polish physicist Marie Curie and her husband, the French chemist Pierre Curie. While studying a sample of the uranium ore pitchblende, the Curies noticed that the level of radioactivity it contained was four times what was expected. Having analysed the ore and searched for all the known elements, they surmised that there must be at least one unknown element present. In fact, there were two: they discovered polonium first, then radium later the same year. The husband-and-wife team named the first element after Marie's country of birth. Polonium occupies a position in the periodic table that was tentatively suggested by Dmitri Mendeleev in 1871; he called the yet-to-be-discovered element *eka-tellurium* ("below tellurium").

As one would expect of a highly radioactive element, polonium is very rare. The longest-lived naturally occurring isotope, polonium-210, has a half-life of just 138 days, so any polonium that was present in the early Earth has long since disintegrated. However, new polonium atoms are created constantly inside uranium ores, as part of a decay chain that begins with the disintegration of uranium atoms. A sample of polonium-210 produces noticeable heat as a result of the energy released by radioactive decay; it was once used as a source of power in thermoelectric batteries used in spacecraft. The longest-lived isotope of polonium (polonium-209), created artificially, has a half-life of a little over 100 years.

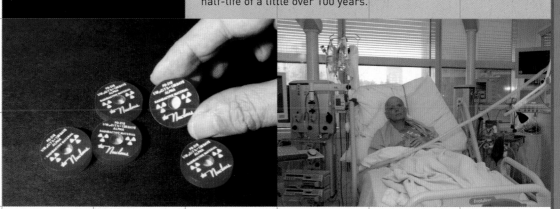

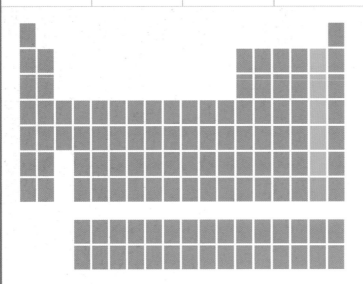

9	
F	
Fluorine	

17	
Cl	
Chlorine	

35	
Br	
Bromine	

53	
I	
Iodine	

85 ☢
At
Astatine

117 ☢
Uus
Ununseptium

The Halogens

Unlike Groups 14, 15 and 16 of the periodic table, Group 17 is populated by elements that have very similar properties. They are all highly reactive non-metals that form singly charged negative ions and form similar compounds. Nevertheless, there are differences when these elements are pure: fluorine and chlorine are gases at room temperature, bromine is liquid and iodine is solid. These differences are mainly the result of increasing atomic mass down the group. Astatine also stands apart – it is very unstable and highly radioactive. This group also includes element 117, which has the temporary name "ununseptium" (1-1-7-ium). Element 117 does not occur naturally, but has been created in nuclear laboratories; it features with the other transuranium elements, on pages 153–7.

The electron configuration for each of these elements ends with $s^2 p^5$; one more electron and the p-orbitals would be full, so atoms of these elements are one electron short of a completely filled shell. The Group 17 elements can be considered the opposite of the Group 1 elements (see page 22): atoms of Group 1 elements easily lose an electron to attain a filled shell of electrons, while atoms of Group 17 elements easily gain one. As a result, all of the elements of Group 17 most often occur as negative ions (anions) in ionic compounds – commonly with positive ions of the alkali metals. An example is common salt, sodium chloride (NaCl), which is composed of positively charged sodium ions and negatively charged chloride ions.

Group 17 elements are called halogens – a term invented by Swedish chemist Jöns Jacob Berzelius in 1842, although "halogen" had been previously suggested for the element now known as chlorine. The word "halogen" is derived from the Greek word for salt, hals, because the first four elements all react with metals to produce compounds similar to common salt.

9

F

Fluorine

ATOMIC NUMBER: 9

ATOMIC RADIUS: 50 pm

OXIDATION STATES: -1

ATOMIC WEIGHT: 19.00

MELTING POINT: -220°C (-363°F)

BOILING POINT: -188°C (-307°F)

DENSITY: 1.70 g/L

ELECTRON CONFIGURATION: [He] $2s^2 2p^5$

Like all the halogens (Group 17 elements), the element fluorine is very reactive in its elemental form, due to the fact that it is one electron short of a filled outer electron shell (in fluorine's case, the second shell, n=2; see pages 8–9). Fluorine atoms easily accept an extra electron from other atoms, and as a result, they become negatively charged fluoride ions. While fluorine itself is reactive, fluoride ions (F^-) are very stable because of their filled outer electron shell ($2s^2 2p^6$).

The fact that fluorine atoms are one electron short of a filled electron shell – coupled with the fact that they are small, having only nine electrons altogether, all held tightly to the nucleus – means that fluorine atoms very rarely form covalent bonds (see pages 11–12). They have a tendency to "steal" the electrons involved in covalent bonds for themselves – a property called electronegativity. Fluorine is the most electronegative of all elements. One exception is when fluorine atoms bond with other fluorine atoms, forming two-atom covalently bonded molecules, F_2. The diatomic F_2 molecule is fluorine's normal elemental form; the same is true of the other halogens (and also the elements hydrogen, nitrogen and oxygen). Elemental fluorine is a pale yellow gas at normal temperatures.

Fluorine is the thirteenth most abundant element in Earth's crust, where it exists almost entirely in the form of fluoride ions in minerals. The most important fluoride-containing mineral is fluorite, also known as fluorspar, which consists mostly of calcium fluoride (CaF_2). From the sixteenth century, fluorspar was added to iron ores, lowering the temperature at which the metals melt when the ores were smelted, and making the whole mixture flow. Fluorspar (fluorite) is still used in steelmaking today, for the same reason. The Latin term for the word "flow" is *fluo*, and from this comes the name "fluorspar" and, once the element had been identified, the name "fluorine". Another common English word, "fluorescence", is derived from the word "fluorspar". Fluorescence is the property whereby some substances glow in ultraviolet light. English physicist George Gabriel Stokes coined the term in 1852, with the meaning "to assume the state of fluorspar"; the mineral was well known for its fluorescence.

It was French physicist and chemist André Marie Ampère who first suggested the existence of an unknown element in "fluoric acid" made from fluorspar (the acid is now known as hydrofluoric acid). English chemist Humphry Davy suggested the name "fluorine" in 1813, but it was more than 70 years before anyone could produce fluorine gas, because of fluorine's extreme reactivity and the resulting stability of its compounds. It was French chemist Henri Moissan who finally succeeded, in 1886.

Below: Crystal of the mineral fluorite, also known as fluorspar, which is composed mostly of calcium fluoride (CaF_2).

To this day, only relatively small amounts of elemental fluorine are produced (less than 20,000 tonnes worldwide). Industrially produced fluorine gas is mostly reacted almost immediately to form useful compounds, since the element is difficult and dangerous to store and transport. The main uses for elemental fluorine are for producing sulfur hexafluoride (SF_6), used in heavy-duty electrical switches (see page 129), and for producing uranium hexafluoride (UF_6), used in the enrichment of uranium (see page 97).

The starting point for most of the varied industrial uses of fluorine – including fluorine gas – is hydrogen fluoride (HF), which is known as hydrofluoric acid when in solution. Hydrofluoric acid is produced by reacting fluorite (fluorspar) with sulfuric acid. It is used to make a number of organic (carbon-based) compounds, often with other halogens. The best known example is a class of compounds called CFCs (chlorofluorocarbons). CFCs were first produced in the nineteenth century, but from the 1920s they were used as a refrigerant in domestic and industrial refrigerators and as propellants in aerosol cans. In the 1980s, it was found that when these gases are released into the atmosphere, they cause the slow, steady destruction of the protective ozone layer (see page 124). The use of CFCs in consumer products was banned by the Montreal Protocol on Substances that Deplete the Ozone Layer,

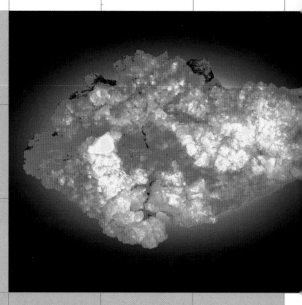

which came into force in 1989. Today, the related classes of compounds HFCs (hydrofluorocarbons) and HCFCs (hydrochlorofluorocarbons) are the most widely used refrigerants. Another widely used organic fluoride compound is polytetrafluoroethylene (PTFE); its main applications are as an insulator in electrical wiring; as white, waxy, waterproof tape used by plumbers; and, in an "expanded" form, as the breathable fibres in Gore-Tex® jackets. Known by its trade name Teflon®, PTFE is also used as the coating of non-stick cooking pans.

Hydrofluoric acid is used to make a large number of inorganic (not carbon-based) compounds. One of the most important is sodium hexafluoroaluminate (Na_3AlF_6), which is used molten as a solvent in the extraction of aluminium. When the modern process for aluminium extraction was pioneered in the 1880s (see page 101), it made use of the mineral cryolite, which is mostly sodium hexafluoroaluminate. But natural cryolite is rare; today "synthetic cryolite", made from hydrofluoric acid, accounts for around one-third of all the hydrofluoric acid produced.

Above: The mineral fluorite glowing – fluorescing – under illumination by ultraviolet radiation. The word "fluorescence" is derived from the name of this mineral.

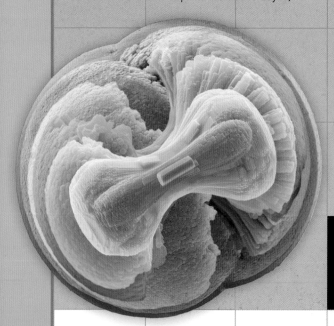

Left: Coloured scanning electron micrograph (magnification x1800) of a laboratory-grown crystal of the mineral fluorapatite, which contains fluoride ions. It is this durable mineral that forms in teeth as a result of fluoride in water and toothpaste.

Hydrofluoric acid is also used to produce sodium fluoride (NaF) and sodium monofluorophosphate (Na_2PO_3F), fluoride-containing compounds added to toothpastes. Fluoride's main role in toothpaste is to re-mineralize the tooth enamel from the calcium mineral hydroxylapatite to the fluoride mineral fluorapatite, which is more resistant to decay. Fluoride ions are present in small amounts in most water. In some areas, the natural fluoridation in water supplies has a beneficial effect on teeth; in areas with very little fluoride present, some water companies add fluoride to water supplies in an effort to improve dental health. Some foods also contain fluoride – most notably, a single cup of tea typically contains 0.5 mg (milligrams), but can contain more than 1 mg. Typical overall daily fluoride intake is around 3 mg.

In certain locations, the natural level of fluoride in water is much higher than normal. Chronic excessive intake of fluoride ions can lead to dental fluorosis, a disease characterized by the weakening of tooth enamel. In the worst cases, the damage can extend to bones. In skeletal fluorosis, limbs may become deformed as well as more prone to fracture. Excessive fluoride intake has other detrimental effects on human health, including kidney damage and hypothyroidism (under-activity of the thyroid gland). As a result, the fluoridation of water supplies is a highly controversial issue; those who protest against it consider it a widespread toxin, while those in favour cite studies that seem to show its beneficial effects in reducing the rate of tooth decay.

Elemental fluorine is highly toxic, and fluoride ions are moderately so – although ingesting more than about 5 grams (5000 mg) of fluoride in a single dose is normally fatal. Fluoride compounds have been used as rat poisons and insecticides, and an organic fluoride compound, commonly known as sarin, is a potent nerve gas that was once used in chemical warfare but whose production has now been outlawed.

Above: Nuclear isotope separation plant. Inside, gaseous uranium hexafluoride (UF_6), prepared from uranium ore, is separated into those molecules containing the isotopes uranium-238 and those containing uranium-235; the latter are used in fission reactors.

Left: Cycling shoes made with the water-resistant yet breathable synthetic fabric Gore-Tex®, a porous form of the polymer polytetrafluoroethylene.

17

Cl

Chlorine

ATOMIC NUMBER: 17

ATOMIC RADIUS: 100 pm

OXIDATION STATES: **-1**, +1, +2, +3, +4, **+5**, +6, **+7**

ATOMIC WEIGHT: 35.45

MELTING POINT: -102°C (-151°F)

BOILING POINT: -34°C (-29°F)

DENSITY: 3.20 g/L

ELECTRON CONFIGURATION: [Ne] $3s^2\,3p^5$

Like fluorine, the element chlorine is a reactive and toxic gas in its elemental form at normal temperatures (it condenses to form a liquid at −34°C). Also like fluorine, elemental chlorine is normally composed of diatomic molecules (Cl_2). It has a distinctive sharp smell, and a distinctive yellow-green tinge to it; the element's name is derived from the Greek word *chloros*, meaning "greenish yellow". Again like fluorine, the gain of a single electron completes its atoms' outer electron shells, so that they attain a stable configuration. As a result, chlorine is normally found in the form of negatively charged ions, called chloride ions.

Chlorine is not quite as reactive as fluorine, and as a result it is easier to liberate from its compounds than fluorine. It is no surprise, then, that chlorine gas was produced more than a hundred years before fluorine gas. It was Swedish chemist Carl Wilhelm Scheele who first produced elemental chlorine when he reacted hydrochloric acid (a solution of hydrogen chloride, HCl) with manganese(IV) oxide (MnO_2). Scheele noted that the gas produced by the reaction had a pungent smell and reacted with metals. It was more than 30 years before chemists realized that what Scheele had produced was a chemical element.

Chlorine gas, produced from brine (salty water, NaCl and H_2O), is now used to manufacture a range of organic chlorine compounds, including CFCs (chlorofluorocarbons; see page 136) and trichloromethane ($CHCl_3$), better known as chloroform. Once popular as an anaesthetic, chloroform is now used mainly in the production of polytetrafluoroethylene (PTFE; see page 136).

Alchemists and early chemists alike had been using hydrochloric acid since the sixteenth century. Today, hydrochloric acid is produced industrially by dissolving hydrogen chloride in water. The hydrogen chloride is made by reacting hydrogen gas and chlorine gas; both are produced by passing electric current through brine. Millions of tonnes of hydrochloric acid are produced each year, since it has many uses in the chemical industry. For example, it is used in the production of PVC (polyvinyl chloride), which has many applications, including plastic door and window frames. Very strong hydrochloric acid is naturally present in the human stomach, where it kills bacteria and aids the digestion of proteins.

Although fluorine is the most electronegative element – it has the highest tendency to "steal" electrons from covalent bonds – chlorine has a higher affinity for electrons. This means that chlorine atoms more easily form negative ions than do fluorine atoms, by stealing electrons from atoms of other elements. The loss of electrons – in this case, from elements with which

Above The element chlorine – a gas at normal pressure and temperature.

Right Nineteenth-century anaesthesia face mask and bottles of chloroform, which was used as an anaesthetic in surgery from 1847 until the 1950s, when it was superseded by less toxic compounds.

Above: The top of a bottle of household bleach. The active bleaching compound is sodium hypochlorite, which decomposes in water to produce highly oxidizing elemental chlorine.

Below: Soldiers and a horse wearing gas masks during the First World War, to prevent inhaling mustard gas, a highly poisonous chemical weapon. The masks also incorporated eye protection, because mustard gas caused conjunctivitis and temporary blindness.

that can be used in the home. Domestic bleach, for example, is a solution of the compound sodium hypochlorite ($NaClO$). The chlorine atoms in the hypochlorite ion (ClO^-) effectively have one less electron than normal (they have an oxidation number of +1). When hypochlorite ions react with other substances, the chlorine atoms take on two electrons each, so that they end up as chloride ions (Cl^-). The removal of two electrons by each chlorine atom is the reason why sodium hypochlorite is such a powerful oxidizing agent, bleach and disinfectant. Hypochlorite compounds are commonly used in water supplies to inhibit the spread of waterborne pathogens. They are also used in swimming pools for the same purpose. Another chlorine compound, chlorine dioxide (ClO_2), is commonly used to bleach wood pulp that is made into paper.

Chlorine's powerful oxidizing ability makes it toxic as well as useful. In the First World War, the German army released chlorine gas into the British trenches, killing and maiming thousands. Chlorine is also a key component of mustard gas – any of a series of organic (carbon-based) sulfur- and chlorine-containing compounds that cause blistering on skin and in the lungs. Mustard gas was also first used in the First World War, and has been used in several other conflicts since then. Within the mustard gas molecule, chlorine removes an electron, breaking away to form a chloride ion and leaving the molecule primed as a reactive ion that attacks DNA. Another chlorine-containing

chlorine comes into contact – is known as oxidation. The origin of the term concerns reaction with oxygen, which also has a high electron affinity. So, for example, when iron atoms come into contact with oxygen, the oxygen removes electrons from the iron, leaving the metal oxidized – in other words, the iron rusts. Chlorine's high electron affinity makes it a very powerful oxidizing agent, and this is the property behind most of its many applications.

Oxidation often results in bleaching, since the reactions involved can break crucial bonds in pigment molecules. When Scheele investigated the gas he had produced, he noted that it could bleach leaves, for example. Oxidizing agents can also destroy pathogens such as bacteria and viruses, and chlorine can be used as a disinfectant; chlorine oxidizes compounds in the cell membranes of bacteria, for instance, making the bacterial cells spill their contents and leading to their death.

While elemental chlorine is a powerful oxidizing agent, and a powerful bleach and disinfectant, some compounds of chlorine are more powerful still – and, unlike chlorine, they are available in a form

compound notorious for its use as a suffocating chemical weapon in the First World War is phosgene (carbon oxychloride, $COCl_2$). Today, it has several uses in the chemical industry, including an important role in the production of polycarbonate, which is used to make spectacle lenses, plastic drinks bottles, and CDs, DVDs and Blu-ray™ discs.

During the twentieth century, many other organic chlorine-containing chemicals were developed for peaceful reasons. For example, each molecule of the insecticide DDT (dichlorodiphenyltrichloroethane) contains five chlorine atoms. DDT was widely used in the 1940s and 50s, but its use was curtailed and then banned after American biologist Rachel Carson found that DDT was carcinogenic, and was also having a devastating effect on the food chain. In particular, she found that the indiscriminate use of DDT was dramatically reducing the populations of songbirds; Carson detailed her findings in her influential 1962 book *Silent Spring*, considered by many as the beginning of the modern environmental movement.

Another class of organochlorine compounds are PCBs (polychlorinated biphenyls), which were once widely used as coolants in electrical switchgear and transformers. A 2001 United Nations treaty, called the Stockholm Convention on Persistent Organic Pollutants, restricts the production and use of a range of organic compounds that persist in the environment and have undesirable effects on wildlife and human health. Nearly all the substances restricted by the treaty contain chlorine; the list includes DDT and PCBs.

Even the chlorine in our drinking water and swimming pools can pose problems to human health. The reaction of chlorine with organic matter such as skin cells and sweat produces a range of what scientists call disinfection by-products, some of which are genotoxic – they can cause mutations in DNA that potentially lead to cancer. However, the risk is low, and the health benefits of swimming in water free of pathogens far outweigh the potential risk. Sea water contains much more chlorine than swimming-pool water – but it is in the form of chloride ions, which are stable and are not reactive or toxic in the same way as chlorine in hypochlorite or other highly oxidizing chlorine compounds.

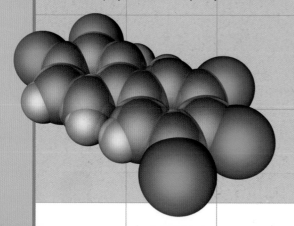

Top left: Vinyl (PVC) record – the most popular format for bought music before the advent of CDs and MP3 players. PVC is also used to make plumbing waste pipes, guttering and plastic door and window frames.

Top right: Swimming pool, in which dissolved hypochlorite compounds decompose, releasing elemental chlorine into the water. Dissolved chlorine is very effective at killing potentially harmful microbes.

Left: Molecule of one of the class of toxic compounds called polychlorinated biphenyls. This molecule, $C_{12}H_5Cl_5$, consists of two rings of carbon atoms (dark grey), five hydrogen atoms (white) and five chlorine atoms (green).

35

Br

Bromine

ATOMIC NUMBER: 35
ATOMIC RADIUS: 115 pm
OXIDATION STATES: **-1**, **+1**, **+3**, +4, **+5**, +7
ATOMIC WEIGHT: 79.90
MELTING POINT: -7°C (19°F)
BOILING POINT: 59°C (138°F)
DENSITY: 3.10 g/cm³
ELECTRON CONFIGURATION: [Ar] $3d^{10} 4s^2 4p^5$

The red-brown element bromine is one of only two elements that are liquid at normal room temperature (the other is mercury). It is dense: a litre of bromine would weigh the same as more than three litres of water. In its elemental form, it exists as diatomic molecules, Br_2. Like all the halogens (Group 17 elements), bromine is a non-metal – although under high pressures, it does become a metalloid, conducting electricity and heat fairly well, and taking on a metallic lustre.

Right: Flask containing elemental bromine – a red-brown liquid at normal pressure and temperature, but so volatile that it is always giving off fumes.

Bromine occurs naturally in compounds in Earth's crust, where it is relatively rare: it is the sixty-second most abundant element. Because it dissolves in water, as bromide ions (Br⁻), the world's oceans, lakes and rivers contain billions of tonnes of it. There is no known biological role for bromine in humans, but some marine organisms do absorb small amounts of it, for reasons that are not clear. Certain species of sea snail produce a compound that was used to make a dye called Tyrian purple, which was highly prized in Ancient Phoenicia (modern day Lebanon, Syria and Israel).

Bromine boils at 59°C – but even at room temperature, it readily evaporates to form a visible and pungent vapour. The name of the element is derived from the Greek word *bromos*, meaning "stench". French apothecary (pharmacist) and part-time chemistry teacher Antoine-Jérôme Balard extracted an unknown red-brown liquid in 1825, in the briny liquid from a salt marsh. At first he thought the liquid must be a compound of iodine and chlorine, but he soon became convinced that it was a new element. Also in 1825, German chemistry student Carl Löwig independently produced the same mysterious red-brown liquid, from mineral water near his home. Balard's account was published first, so he is normally credited with the discovery. One of the first major applications of bromine, from 1840, was in Daguerreotype photography (see page 75). In 1857, English physician Charles Locock reported that he had prescribed potassium bromide (KBr) to female patients suffering from epilepsy during menstruation, with great success. Gradually, the compound was used to treat epilepsy that was not associated with menstruation, in men as well as women. Potassium bromide was the first successful anti-epileptic drug, and remained the only one until the introduction of the drug phenobarbital in 1912.

Bromine is a strong oxidizing agent – and, like chlorine, its compounds are sometimes used in water treatment, particularly in hot spa baths. But it is not as strong an oxidizer as chlorine, a fact that is put to use in extracting the element from sea and lake water. Chlorine gas is bubbled through the water, and chlorine atoms steal electrons from the bromide ions in solution, oxidizing them to elemental bromine. A few hundred thousand tonnes of bromine are produced in this way worldwide each year. Most of it is used to make very effective flame retardants that are mixed into plastics and textiles. Brominated flame retardants are commonplace in industry, and until recently, they were used in many consumer products. But concerns over bromine's toxicity and its effects on the ozone layer have led many manufacturers to seek alternative compounds. Several other historic uses of bromine compounds – from sedatives to insecticides – were phased out because of the same concerns, or because better alternatives were developed, but many small industrial applications remain. Bromine compounds are sometimes used as fluids to aid in drilling for oil, and some organic bromine compounds are still used as dyes.

53

I

Iodine

ATOMIC NUMBER:	53
ATOMIC RADIUS:	140 pm
OXIDATION STATES:	**-1**, **+1**, **+3**, +5, +7
ATOMIC WEIGHT:	126.90
MELTING POINT:	114°C (237°F)
BOILING POINT:	184°C (364°F)
DENSITY:	4.94 g/cm³
ELECTRON CONFIGURATION:	[Kr] 4d¹⁰ 5s² 5p⁵

Right: Conical flask containing elemental iodine – a deep purple solid at normal pressure and temperature. Iodine sublimes (turns from a solid directly into a gas).

Below: Bladder wrack (*Fucus vesiculosus*) seaweed, a good source of iodine that was popular in the nineteenth century for treating goitre, a swelling of the thyroid gland caused by iodine deficiency.

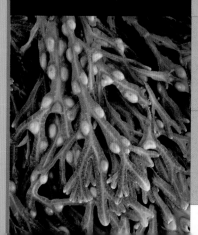

In its elemental form, iodine exists as a solid composed of diatomic molecules (I_2). It has a deep blue-black colour with an almost metallic sheen. Even at room temperature, countless iodine molecules escape, breaking away from the solid to form a deep violet vapour; this production of vapour directly from a solid is called sublimation. Iodine does form a thick brownish liquid, at 114°C. It is not as reactive as fluorine, chlorine or bromine – but it, too, forms compounds with many elements. Like the other halogens (Group 17 elements), iodine easily forms stable negative ions, called iodide ions (I^-), so most iodine compounds are ionic.

Atoms and ions of iodine are fairly rare in the Universe and on Earth, but there are several minerals that contain iodide ions. Much of the ten or so thousand tonnes produced each year comes from these minerals, the rest being extracted from brine. Iodine itself is only sparingly soluble in water, but iodide ions from ionic compounds dissolve readily. As a result, seawater contains a small amount of iodide, and in briny waters, the concentration is higher. Iodine occurs in organic (carbon-based) compounds in certain marine organisms – particularly as methyl iodide (CH_3I), and particularly in seaweed and some other algae.

Iodine was discovered in 1811, by French chemist Bernard Courtois, who was trying to extract sodium and potassium compounds from seaweed as part of his effort to produce potassium nitrate for the production of gunpowder (see page 128). When Courtois added sulfuric acid to the ashes left behind after burning seaweed, a purple vapour was produced, which condensed to form curious dark blue crystals. Other chemists carried out further investigations, and announced that Courtois had found a new element. French chemist Joseph Gay-Lussac gave it the name *iode*, from the Greek *iodes*, meaning "violet".

As early as 1829, French physician Jean Lugol made up a solution of iodine and iodide that he believed could treat tuberculosis. He was wrong, but the solution is still used today – as is a similar solution of iodine and iodide called tincture of iodine – as an antiseptic, in water purification and as a laboratory test for the presence of starch. When iodine is dripped onto starchy foods, it stains the starch dark purple. Like bromine (see page 141), iodine was used in the Daguerreotype process in early photography; iodine vapour reacted with silver to form silver iodide, which is light-sensitive. Silver iodide is very important for this reason in film-based photography. It is also used in cloud seeding: crystals of the compound are sprinkled into clouds to stimulate the fall of rain.

Iodine is essential in humans – although the total amount of the element present in the body is only around 20 milligrams. Its most important role is in synthesis of two hormones produced in the thyroid. The hormones – triiodothyronine and thyroxine – have many crucial roles, including regulation of metabolism, body temperature, heart rate and growth. Too much iodine in the diet can lead to a condition called hyperthyroidism, in which the body produces excess thyroid hormones, leading to an increased metabolic rate and hyperactivity. Deficiency of iodine in the diet leads to hypothyroidism, a condition in which insufficient thyroid hormones are produced. This has many symptoms, including fatigue, weight gain and the intolerance of cold – and in advanced stages, can lead to the development of a goitre, which is a large swelling in the neck. In infants, chronic iodine deficiency can lead to stunted physical growth and impaired mental development (cretinism). Hundreds of millions of people worldwide are at risk from or suffer from iodine deficiency, and much of the cooking and table salt produced around the world has iodide added, in an attempt to counteract the problem.

Radioactive iodine-131 is a common product of nuclear fission reactions (see page 96). It is released into the air as a result of nuclear weapon tests and accidents at nuclear reactors. Although it has a short half-life of just over eight days, it can end up in products such as milk consumed by humans. To prevent this radioactive isotope from being absorbed into the thyroid, public health authorities typically distribute tablets containing potassium iodide (KI), which saturate the thyroid with non-radioactive iodine, to prevent the take-up of the radioactive isotope.

Above: Patient with a large goitre. Iodine deficiency is the cause of more than 90 per cent of cases of goitre worldwide. It is most prevalent in areas where heavy rainfall or flooding has leached away iodine from the soil.

85   ☢

At

Astatine

ATOMIC NUMBER: 85

ATOMIC RADIUS: 125 pm

OXIDATION STATES: **-1**, +1, +3, +5

ATOMIC WEIGHT: (210)

MELTING POINT: 302ºC (576ºF)

BOILING POINT: 335ºC (635ºF), estimated

DENSITY: Not known

ELECTRON CONFIGURATION: $[Xe]\, 4f^{14}\, 5d^{10}\, 6s^2\, 6p^5$

Astatine is a highly radioactive element – any bulk quantity would release so much energy, and produce so much heat, that it would vaporize immediately. If it were not unstable, it would almost certainly share most if not all properties of the halogens (Group 17 elements).

The element astatine exists naturally on Earth only as a result of the decay of other radioactive elements within uranium ores. Only minute quantities exist at any one time – probably less than 30 grams in the entire Earth's crust. It was first identified in 1940, after being created in a laboratory at the University of California, by American physicists Dale Corson and Kenneth MacKenzie and the Italian-American physicist Emilio Segrè; it was produced by bombarding the element bismuth with alpha particles (see pages 9–10). The element's name is derived from the Greek word *astatos*, meaning "unstable". The longest-lived isotope of astatine has a half-life of just eight hours.

The isotope astatine-211, produced in nuclear research facilities, shows promise in the fight against cancer. Incorporated into biocompatible compounds that can be injected directly into tumours, astatine-211 provides a steady source of alpha particles, which are absorbed over a short range by DNA in tumour cells, while doing minimal damage to healthy tissues.

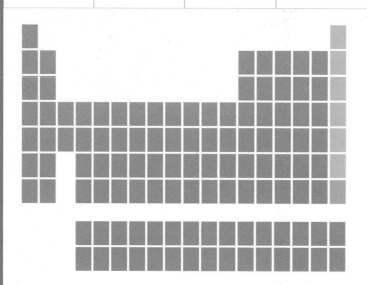

The Noble Gases

The elements of Group 18 of the periodic table – the noble gases – are all very inert (unreactive) gases – although they can be made to form compounds, albeit only with difficulty. Radon, the heaviest naturally occurring noble gas, is highly radioactive. All of the naturally occurring elements of Group 18 were either discovered or first isolated by Scottish chemist William Ramsay. This group also includes element 118, which has the temporary name "ununoctium" (1-1-8-ium). Element 118 does not occur naturally, but has been created in nuclear laboratories; it features with the other transuranium elements, on pages 153–7.

The inertness of the noble gases is explained by the fact that all these elements have filled electron shells; this is also why they appear at the extreme right-hand end of the periodic table. So inert are the atoms of these elements that they do not even bond with their own kind: unlike the halogens of Group 17, which exist as diatomic molecules such as F_2 and Cl_2, the Group 18 elements are all monatomic. They all have very low boiling points: they remain gaseous until they are cooled to very low temperatures. The boiling point is related to the atomic mass, so it increases down the table; helium has the lowest boiling point of any element, at −269°C, while radon boils at −61°C. The atoms of Group 18 elements are spherical – a condition that only applies to atoms that have filled or half-filled orbitals (see also page 114).

2	**He** Helium
10	**Ne** Neon
18	**Ar** Argon
36	**Kr** Krypton
54	**Xe** Xenon
86	**Rn** Radon
118	**Uuo** Ununoctium

2

He

Helium

ATOMIC NUMBER: 2

ATOMIC RADIUS: approx. 130 pm

OXIDATION STATES: (none)

ATOMIC WEIGHT: 4.00

MELTING POINT: -272°C (-457°F), only under high pressure

BOILING POINT: -269°C (-452°F)

DENSITY: 0.18 g/L

ELECTRON CONFIGURATION: $1s^2$

The lightest of the Group 18 elements, helium, is familiar as the lighter-than-air gas that fills party balloons. Like all the noble gases, helium is colourless, odourless and extremely unreactive. There are no known stable compounds of helium; it exists only as stubbornly individual atoms. Helium has the lowest melting and boiling points of any element.

Above right:
Helium discharge lamp in the shape of the element's symbol. Helium is a colourless gas at normal temperatures and pressures. The light is the result of electrons in helium returning to lower energy levels after being excited by the electric current passing through.

Right: False-colour image of the Sun produced by an instrument aboard the Solar Heliospheric Observatory satellite (SOHO). The instrument detects ultraviolet radiation emitted by helium ions in the Sun's chromosphere, a layer normally obscured by the bright outer layer, the photosphere.

Helium was one of just two elements formed in any great quantities during the first few minutes of the Universe (the other was hydrogen). What formed early in the Universe were actually helium nuclei, not helium atoms. There were no atoms of any kind until around 300,000 years after the Big Bang – the "epoch of recombination", when the temperature of the Universe had dropped sufficiently for electrons to combine with nuclei. The other elements were manufactured much, much later, deep in stars and in cataclysmic supernova explosions (see page 8).

The helium nuclei created in the early moments of the Universe included the two stable isotopes of helium: helium-4 (2 protons, 2 neutrons) and helium-3 (2p, 1n). Both these isotopes are still being made inside every shining star, as a result of nuclear reactions in which hydrogen nuclei join (fuse). These fusion reactions are the main source of a star's energy. Since large amounts of helium were created just after the Big Bang and stars are constantly producing more, it is no surprise that this element is the second most abundant in the Universe (after hydrogen). Nearly one in 10 of all the nuclei or atoms in the entire Universe is helium.

Inside larger stars, helium nuclei fuse to make heavier nuclei. For example, each nucleus of the most common isotope of oxygen, oxygen-16 (8p, 8n), is made of four helium-4 nuclei fused together. Some large nuclei are unstable, and one of the ways they attain stability is by downsizing, ejecting a clump made of two protons and two neutrons, known as an alpha particle (see pages 9–10). An alpha particle is identical to a helium-4 nucleus (2p, 2n) – so whenever radioactive substances undergo alpha decay, they create helium-4 nuclei. If and when an alpha particle gains two electrons, it becomes a neutral atom of helium.

In the Universe as a whole, helium-4 nuclei outnumber helium-3 by 10,000 to one. But the ratio of the two isotopes varies considerably, because in certain circumstances, more helium-4 is produced than helium-3. Here on Earth, for example, where radioactive decay creates helium-4, the ratio is more like one million to one.

Oddly perhaps, for an element that is so abundant in the Universe at large, there is relatively little helium on Earth – particularly in the atmosphere. What helium there is in the air is concentrated in the very outer reaches of the atmosphere. Overall, only one in every 200,000 or so particles of gas that make up the atmosphere is a helium atom. Unlike, say, nitrogen, which makes up 78 per cent of

the atmosphere, most helium atoms escaped into space early in Earth's history. The less massive the particles of a gas, the greater is their average speed; light individual helium atoms (He) travel much faster on average than heavier diatomic nitrogen molecules (N_2), for example. Some helium atoms move fast enough to escape Earth's gravity; those that didn't escape straight away did eventually attain high enough speeds, as the result of collisions with other atoms and molecules. By contrast, the gas giant planet Jupiter has retained most of its helium; its dense atmosphere contains around 24 per cent helium by mass. This is because the planet is colder (being further from the Sun) and has a much stronger gravitational pull (being more massive).

As a result of its scarcity in Earth's atmosphere, helium for use in party balloons and its many other applications is extracted not from the air, but from underground sources. It is a by-product of the natural gas industry: the mixture of gases withdrawn from natural gas wells typically contain a few per cent helium. This underground helium is constantly being produced by the decay of radioactive elements in Earth's crust; several thousand tonnes of helium are created in this way each year. Some of it becomes trapped in impermeable rock formations – the same traps in which natural gas accumulates.

Helium's rarity in Earth's atmosphere, together with its extreme inertness, meant that its discovery came after most of the other naturally occurring elements had been discovered. In 1868, French physicist Pierre Jules Janssen and English physicist Joseph Norman Lockyer independently noticed a bright line in the spectrum of sunlight that did not correspond to any of the elements known at the time. Both scientists suggested that the line must therefore belong to an unfamiliar element. Lockyer suggested the name "helium" for

the element, derived from the Greek word for the Sun, *helios*. Some scientists mocked the idea that an element could be present in the Sun but not on Earth. In 1895, Scottish chemist William Ramsay isolated a gas given off when he mixed a uranium ore with acids. Ramsay was searching for argon, which he had isolated the previous year, but when he analysed the spectrum of the gas, he realized that it was Lockyer's helium. Ramsay sent a sample of the gas to Lockyer, who was overjoyed to see the same "glorious yellow effulgence" he had observed in the spectrum of sunlight nearly 30 years earlier.

In the next 20 years, only small amounts of helium were isolated, and scientists thought that helium was extremely rare on Earth. A surprise discovery in 1905 changed all that. Two years earlier, a large crowd gathered in the town of Dexter, in Kansas, USA, to watch what should have been the dramatic lighting of a stream of gas coming from a newly discovered natural gas well. Burning straw bales were repeatedly pushed on to the gas stream, but the gas would not light, much to the embarrassment of local officials. Analysis of the gas showed that it was nearly three-quarters inert nitrogen, but further analysis revealed that nearly 2 per cent of the gas was helium. As a result of the discovery that helium is present in many natural gas deposits, the price of helium plummeted. By 1924, the USA had built a facility to stockpile the gas, mostly for use in airships. Today, that same facility – the US National Helium Reserve – contains more than a billion cubic metres of helium. The USA produces the majority of the world's helium, Algeria most of the rest.

Helium gas is a very good conductor of heat, and it is used as a coolant in some gas-cooled nuclear reactors. Helium's inertness makes it useful as a protective atmosphere in which to manufacture

Left: Helium-filled airship awaiting lift-off. Helium is less dense than air, so gives the craft buoyancy.

Below: Eye surgery with a helium-neon laser. This type of laser was long used in barcode readers and CD players but has been superseded by smaller, cheaper diode lasers in those applications.

very sensitive materials or components. In the electronics industry, crystals of ultra-pure silicon are grown, and liquid crystal displays (LCD) and optical fibres are manufactured, in helium atmospheres. Helium's inertness also makes it useful as a shielding gas in arc welding; the helium prevents oxygen from the air reaching the intense heat of the weld, where it would oxidize the metals involved.

Several modern applications of helium involve the element being used in liquid form. Liquid helium is used to cool superconducting magnets used for MRI (magnetic resonance imaging), for example. Similar but much bigger superconducting magnets in particle accelerators also depend upon liquid helium, and gaseous helium is used to remove heat produced by particle collisions. In space technology, liquid helium is used to liquefy hydrogen gas, for use as a rocket fuel – and to liquefy oxygen, with which the fuel can burn – and to keep them cool prior to launch. Many components of spacecraft and aeroplanes are treated to thermal cycling during manufacture – a process in which repeated cooling and heating makes the components hardwearing and stable enough for the demands put on them in use. The cold phase of each thermal cycle is normally achieved using liquid helium.

A common trick that people play with helium is to inhale the gas from party balloons, for fun, to make their voice sound squeaky. Contrary to popular belief, breathing helium does not change the *pitch* of your voice: the vocal folds in a person's voice box produce exactly the same range of frequencies whether in air or in helium. But the speed of sound is much higher in helium, and this has the same effect that shrinking the cavities inside the mouth and nose would have, so that higher frequencies resonate. Taking a single breath, as a joke, does no harm, as helium is non-toxic – but breathing it more continuously can be dangerous and has caused deaths. With lungs full of helium (or any other gas), oxygen is prevented from reaching the body's tissues, and a person can suffocate.

Above: One of the cryogenics plants supplying liquid helium to the Large Hadron Collider, the world's largest particle accelerator. The accelerator is a 27-kilometre ring in which protons are guided by superconducting magnets kept cool by 100 tonnes of liquid helium.

10

Ne

Neon

ATOMIC NUMBER: 10

ATOMIC RADIUS: approx. 160 pm

OXIDATION STATES: (none)

ATOMIC WEIGHT: 20.18

MELTING POINT: -249°C (-415°F)

BOILING POINT: -246°C (-411°F)

DENSITY: 0.90 g/L

ELECTRON CONFIGURATION: [He] $2s^2\,2p^6$

As for all the noble gases, the element neon is colourless, odourless and very inert (unreactive). Like helium, neon forms no known stable compounds. The name "neon" is derived from the Greek word *neos*, meaning "new". Neon is the fourth most abundant element in the Universe overall, but is very rare on Earth. It accounts for only about 0.002 per cent of the atmosphere, making it slightly more abundant than helium.

Scottish chemist William Ramsay and English chemist Morris Travers discovered neon in a sample of liquefied air in 1898, a few weeks after discovering krypton (see page 150). The two chemists separated off the part of the air they knew was argon, and allowed that to warm separately, to see if another element was present. When

Above: Neon discharge lamp in the shape of the element's symbol. Neon is a colourless gas at normal temperatures and pressures. The light is the result of electrons in neon returning to lower energy levels after being excited by the electric current passing through the tube.

they passed an electric current through the gas given off, they were surprised to see the gas emit a bright red-orange glow. Analysis of the spectrum of the light confirmed that it was a previously unknown element.

In 1910, French inventor Georges Claude found a way to make use of the bright glow produced when electric current passes through rarified (low-pressure) neon gas: he invented the neon tube. Claude's invention became commonplace in advertising, since long neon tubes could be shaped into eye-catching product names and slogans. Neon signs are still common, although today's "neon" tubes contain various mixtures of noble gases, and sometimes mercury vapour, too. There are also many alternative ways of producing eye-catching lighting effects, so the use of true neon signs is declining.

Smaller, lower-voltage neon lamps that work on the same principle as Claude's neon tubes were commonplace as on-off indicator lamps in electronics before the invention of the LED (light-emitting diode) in the 1970s. Flat screen plasma displays are an offshoot of neon lamp technology; these displays are made of millions of tiny sealed capsules, or cells, containing neon, as well as xenon and sometimes krypton. When a voltage is applied across the cell, the gases are ionized and an electric current flows through them. The gas atoms produce ultraviolet radiation. The interior of each cell is coated with either a red, green or blue phosphor paint, which glows when the radiation hits it. Each set of red, green and blue cells forms one pixel of the display. Because of its scarcity and the resulting high cost, neon has few practical applications beyond lighting and plasma displays.

Above: Illuminated signs in Shanghai, China, including neon signs in the shape of Chinese characters. Many "neon" signs are actually filled with a mixture of noble gases, to achieve different colours.

18

Ar

Argon

ATOMIC NUMBER: 18

ATOMIC RADIUS: approx. 175 pm

OXIDATION STATES: (none)

ATOMIC WEIGHT: 39.95

MELTING POINT: -189°C (-309°F)

BOILING POINT: -185°C (-303°F)

DENSITY: 1.78 g/L

ELECTRON CONFIGURATION: [Ne] $3s^2\,3p^6$

Like the other Group 18 elements, argon is an odourless, colourless noble gas that exists as individual atoms, rather than as diatomic molecules like oxygen (O_2) or nitrogen (N_2); it is also extremely inert (unreactive), and has been made to form compounds only under extreme conditions. The element's name is derived from the Greek word *argos*, which means "lazy" or "inactive".

The element argon accounts for nearly 1 per cent of dry air (dry air is used as a standard for the composition of air, as the concentration of water vapour present varies). The first clue to the existence of argon came as long ago as 1785. English chemist Henry Cavendish passed an electric spark through air, in an attempt to determine the

Above: Discharge tube containing the colourless gas argon, glowing as a result of electric current passing through the tube.

percentages of the newly discovered "phlogisticated air" and "dephlogisticated air" (oxygen and nitrogen gases) that comprised it. Nearly 1 per cent of the air remained unaccounted for; we now know that nearly all of that was argon. Nevertheless, it was more than a hundred years before scientists discovered argon as a chemical element. In 1894, English physicist Lord Rayleigh (John William Strutt) noticed that the density of pure nitrogen gas produced by chemical reactions was slightly, but consistently, less than the density of supposed pure nitrogen extracted from the atmosphere. The results suggested that another inert substance was present in the air. Scottish chemist William Ramsay proposed that the substance might be the very same 1 per cent of the atmosphere that Cavendish had found. Working together, Ramsay and Lord Rayleigh were able to isolate argon, determine many of its properties and study its spectrum.

The argon in the atmosphere, amounting to many billions of tonnes, is nearly all argon-40, while the argon in the Sun is largely argon-36. The argon-40 was produced – and is still being produced – as the result of the decay of the radioactive isotope potassium-40. Geologists use this fact to work out the ages of igneous (once molten) rocks – a technique called potassium-argon (K-Ar) dating. Before the rocks solidified, any argon produced could escape, but any produced since then remains trapped in solid rock. By working out how much argon is present in a sample, geologists can work out how long ago the rock solidified. Inside an adult human body, around 400 argon-40 atoms are created every second as a result of the decay of potassium-40 present in food (see page 27).

Argon is by far the most abundant of the noble gases on Earth. It is also relatively easy to produce (from liquefied air), making it also the cheapest of the noble gases. Along with neon, argon is used in "neon" tubes; electrically excited argon produces a bright blue glow. It is also used as an inert filling of incandescent light bulbs – and, mixed with mercury vapour, in fluorescent lamps, which are largely superseding incandescent bulbs.

Like helium, argon is used as a shielding gas in arc welding (see page 147), and as a protective atmosphere for growing pure silicon crystals for the semiconductor industry. Argon also has many applications in metallurgy, including accompanying oxygen in the oxygen lance used in steel converters (see page 124). The flow of argon "stirs" the iron mixture, while the oxygen reacts with carbon and other impurities. Unlike helium, argon is a very poor conductor of heat, and it is used to fill the gap in some double glazing to improve the thermal insulation.

Argon has several important applications in medicine. A procedure known as argon plasma coagulation is used to stop bleeding from tissues inside the body, typically in the large intestine. A stream of argon gas is fed to a probe alongside an endoscope, which allows the surgeon to see what he or she is doing. As it leaves the probe, the argon gas is ionized by a strong electric field; a high-frequency pulsed electric current then passes through the ionized gas, encouraging blood to coagulate and thus aiding healing. Argon is also used in some medical lasers that produce a mixture of precise wavelengths of light in the blue and green parts of the spectrum. These wavelengths are absorbed by cells just beneath the retina and by haemoglobin molecules in the blood. As a result, argon lasers are used to treat leaking retinal blood vessels and detached retinas. They are also used in the treatment of some types of glaucoma. Argon lasers are also used in dentistry to "cure" polymers used as fillings, and in laser light shows.

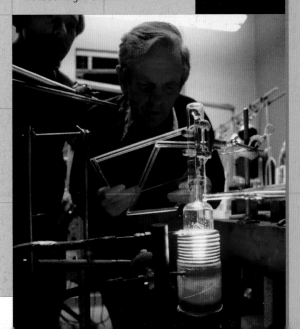

Below: Scientists carrying out potassium-argon dating on volcanic rocks. Introduced in the 1950s, this technique has dated rocks as old as 3.5 billion years and as young as 20,000 years.

36

Kr

Krypton

ATOMIC NUMBER: 36

ATOMIC RADIUS: approx 190 pm

OXIDATION STATES: +2

ATOMIC WEIGHT: 83.80

MELTING POINT: -157°C (-251°F)

BOILING POINT: -153°C (-244°F)

DENSITY: 3.75 g/L

ELECTRON CONFIGURATION: [Ar] $3d^{10}\,4s^2\,4p^6$

The element krypton is a colourless, odourless and very unreactive (inert) noble gas. Because krypton atoms are physically larger than those of helium, neon and argon, the outer electrons are further from the nucleus and they can be made to take part in some reactions. The first krypton compound, krypton(II) fluoride (KrF_2), was produced in 1963; it is a solid, but it is only stable below -78°C.

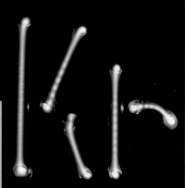

Above: Krypton discharge lamp in the shape of the element's symbol. Krypton is a colourless gas at normal temperatures and pressures. The light is the result of electrons in krypton returning to lower energy levels after being excited by the electric current passing through the tube.

Scottish chemist William Ramsay, with his assistant, the English chemist Morris Travers, discovered krypton in 1898, in a sample of liquefied air. After boiling off the nitrogen, oxygen and argon, they allowed the tiny remaining proportion to boil, and studied the spectrum of light produced when they passed electric current through it. The presence of previously unseen bright yellow and green lines in the spectrum meant that Ramsay and Travers had found a new element. Since they had not yet isolated it, they decided to derive its name from the Greek word *kryptos*, meaning "hidden". Within weeks, the two scientists had discovered neon and xenon in the same sample.

Although it constitutes only around one part per million (0.0001 per cent) of the atmosphere, krypton is obtained industrially from liquefied air. Like neon and argon, it is used in "neon" lights, and also, together with mercury vapour, in fluorescent lamps designed specifically to produce ultra-white light. Krypton-filled arc lamps are installed on airport runways, where they produce short but very bright pulses of light that is visible even through thick fog. Krypton is also used in krypton-fluorine lasers, which are used to etch the microscopic components on computer chips. These lasers contain both krypton and fluorine, which form the compound krypton(II) fluoride for a fraction of a second; when the compound breaks down, it emits a powerful, pure burst of ultraviolet light.

The radioactive isotope krypton-85 is a common product of nuclear fission reactions. It is used to detect leaks in aerospace components. It is also used in some krypton-containing lamps, where the krypton makes it easier to make the lamp start to glow, and, along with argon, in some plasma displays.

The exotic, enigmatic name of element 36 inspired American writer Jerry Siegel and Canadian artist Joe Shuster to invent the planet "Krypton" as the birth planet of their fictional superhero Superman. Planet Krypton first appeared in a comic, together with Superman, in 1938 – 40 years after the element's discovery.

From 1960, the metre (unit of distance) was defined as the combined length of 1,650,763.73 wavelengths of the orange light produced as part of the emission spectrum of the isotope krypton-86. In 1983, that was superseded by the current definition, according to which the metre is the distance light travels in 1/299,792,458 of a second.

54

Xe

Xenon

ATOMIC NUMBER: 54

ATOMIC RADIUS: approx. 220 pm

OXIDATION STATES: +2, +4, +6, +8

ATOMIC WEIGHT: 131.29

MELTING POINT: -112°C (-169°F)

BOILING POINT: -108°C (-163°F)

DENSITY: 5.89 g/L

ELECTRON CONFIGURATION: [Kr] $4d^{10}\ 5s^2\ 5p^6$

Above right: Discharge tube containing the colourless gas xenon, glowing as a result of electric current passing through the tube.

Below: Test firing of a xenon ion engine at NASA's Jet Propulsion Laboratory, California, USA, in preparation for the Deep Space 1 mission.

The heaviest stable element in Group 18, xenon, is a colourless, odourless gas with similar behaviour to the other noble gases. Xenon gas is more than 30 times as dense as helium gas; a xenon-filled balloon is heavier than air and plummets to the ground, while a helium-filled balloon floats upwards. It is even scarcer in Earth's atmosphere than krypton, with concentrations of about 90 parts per billion (0.000009 per cent).

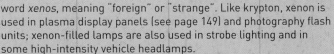

Xenon was discovered in 1898, by Scottish chemist William Ramsay and English chemist Morris Travers, just weeks after the pair had discovered krypton (see opposite). They based the name on the Greek word *xenos*, meaning "foreign" or "strange". Like krypton, xenon is used in plasma display panels (see page 149) and photography flash units; xenon-filled lamps are also used in strobe lighting and in some high-intensity vehicle headlamps.

Xenon was the first noble gas element made to react and form compounds. As is also true for krypton, the relatively large size of the atom means that the outermost electrons are quite far from the atomic nucleus and are more weakly held to the atom. This makes them available for chemical reactions, although only with very electronegative elements, such as fluorine (see page 135) and oxygen. In 1962, English chemist Neil Bartlett produced xenon hexafluoroplatinate ($XePtF_6$), and several other chemists managed to produce several xenon fluorides over the next few years. Today, more than 80 compounds of xenon have been synthesized.

The relatively large size of xenon atoms also means that xenon forms positive ions – by the removal of electrons – more easily than other noble gases. This makes xenon gas the best choice for use in ion engines, advanced technology that can power spacecraft on long missions despite a very low lift-off weight. Inside an ion engine, a stream of electrons create xenon ions, which accelerate to very high speed in a strong electric field. The ions exit the engine as a fast-moving exhaust that pushes the spacecraft forwards ever faster. The first spacecraft to carry an ion thruster was NASA's Deep Space 1, launched in 1998. Deep Space 1 set out with 81.5 kilograms of xenon, and by the end of the mission, after a total of more than 16,000 hours of operation, around 73 kilograms of it had been expelled into space.

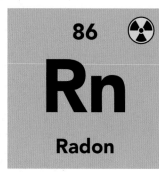

86

Rn

Radon

ATOMIC NUMBER: 86

ATOMIC RADIUS: approx. 230 pm

OXIDATION STATES: +2, +4, +6

ATOMIC WEIGHT: (222)

MELTING POINT: -71°C (-96°F)

BOILING POINT: -62°C (-79°F)

DENSITY: 9.73 g/L

ELECTRON CONFIGURATION: [Xe] 4f^{14} 5d^{10} 6s^2 6p^6

The most dense of the Group 18 elements is another colourless, odourless, unreactive (inert) noble gas – but radon is also highly radioactive. Radon accounts for more than half the natural background radiation on Earth. It is naturally and constantly being created, as a result of the decay of other radioactive elements. The longest-lived and most abundant isotope of radon is radon-222, which has a half-life of about 3.8 days.

Above right: Radon gas detector. Alpha particles emitted during the decay of radon nuclei produce tracks in a plastic film inside the detector. After several months, a count of the tracks reveals the concentration of radon.

Radon gas is produced as part of the decay chains that start with uranium and thorium. It can collect in underground spaces, and can also find its way into the air. In the air – particularly in poorly ventilated confined spaces – radon is a major health hazard for humans. Because it is a gas, the main disease associated with exposure to radon is lung cancer. Miners of uranium are particularly at risk, but so are others in mines that happen to be near to sources of radon. Most of the radioactive dose from radon is actually caused by the decay of its "progeny" – the radioactive isotopes of lead and polonium that are produced when radon itself decays.

Radon was first detected in 1899, when New Zealand-born physicist Ernest Rutherford was studying the radioactivity of thorium. Rutherford found that the activity was greater when the thorium was in a lead box with a door than when it was open to the air, and he concluded that thorium emits an "emanation" that is itself radioactive and retains its activity for a matter of minutes. Other scientists soon found further evidence of Rutherford's emanations. For example, Pierre and Marie Curie found the same thing with radium, but in that case, the emanation remained radioactive for about a month; and in 1903, the same phenomenon was discovered as an emanation from actinium.

In 1904, Scottish chemist William Ramsay, who had already been the first to isolate all the other noble gases, suggested that the emanation from radium might actually be an element, from the family of noble gases, and he suggested that it could be called "exradio" ("from radium"), while the emanations from the other two elements would be "exthoria" and "exactinia". By 1910, Ramsay had collected enough exradio to study many of its properties and to prove that it is a chemical element. He suggested the name "niton", from the Latin word *nitens*, meaning "shining". Nevertheless, until 1923, radon gas was commonly known by three names: radium emanation, thorium emanation and actinium emanation. In that year, the International Union of Pure and Applied Chemistry and the International Committee for Chemical Elements assigned the supposedly three elements the names radon, thoron and actinon – but it gradually became clear that these three distinct entities were isotopes of a single element, and the name radon won through.

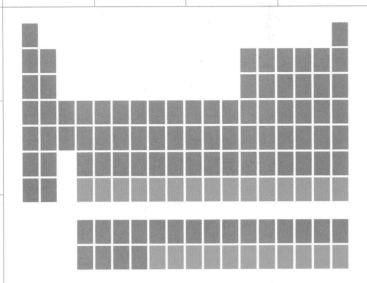

The Transuranium Elements

A transuranium element is any element with an atomic number greater than 92 (the atomic number of uranium). All atoms of a particular element have the same atomic number – the number of protons in the nucleus of its atoms (see page 8) – so every uranium atom has 92 protons in its nucleus. Until the 1930s, scientists supposed that uranium was the heaviest element that could exist. But advances in the understanding and technology of nuclear physics have led to the artificial creation of transuranium elements in laboratories, nuclear reactors, nuclear explosions and particle accelerators.

At present (2016), 26 transuranium elements – with atomic numbers from 93 to 118 – have been created. Six of them (93 to 98) have also been found occurring naturally, in tiny quantities, inside ores of uranium, only after nuclear physicists and chemists had already created them. Several of the transuranium elements, including the first one, neptunium, were produced by bombarding heavy elements with neutrons. Some of the neutrons would be absorbed into the heavy elements' nuclei, making them unstable. The newly-unstable nuclei would undergo beta decay: in which a neutron spontaneously becomes a proton and an electron. Because it now has an extra proton, the nucleus has transmuted into an element with a higher atomic number. Most of the elements with atomic numbers greater than 101 were produced by firing ions or nuclei at heavy elements (see page 156). The transuranium elements from 93 to 103 belong to the actinoid series and are placed in the f-block, which is normally shown separate from the rest of the periodic table (see pages 84–5). The remaining 15 elements are back in the main table, in Groups 4 to 18, period 7.

$$-\frac{\hbar}{i}\frac{\partial}{\partial t} = \frac{p^2}{2m} - \frac{Ze^2}{r}$$

$$\alpha = \frac{\hbar^2}{ec}$$

Above: Italian physicist Enrico Fermi, who with his colleague Emilio Segrè, was the first to try to create transuranium elements, in the 1930s. Fermi was also the first person to build a nuclear reactor – in a squash court at the University of Chicago.

The creation of the first transuranium elements began as the result of the discovery of the neutron in 1932 (see page 17). In 1934, Italian physicist Enrico Fermi bombarded various elements with "slow" (low-energy) neutrons; in nearly every case, the extra neutrons made the elements' nuclei unstable and radioactive. Often, the radioactive nuclei underwent beta decay, in which a nucleus emits an electron. In beta decay, a neutron changes into one proton and one electron ($n \rightarrow p^+ + e^-$). The electron is ejected – to form beta radiation – and the nucleus now has an additional proton. Since the number of protons is the atomic number, the result of beta decay is the "transmutation" of one element into an element with an atomic number *one greater* than before.

Fermi realized what was happening, and he wondered if the same thing might work with uranium – in which case, he would be able to produce element 93, the first transuranium element. He and Italian-born physicist Emilio Segrè tried it – and they tentatively announced that they had been successful. However, they were mistaken; the uranium nuclei had actually split (or fissioned) into two smaller nuclei: Fermi and Segrè had initiated nuclear fission (see page 96).

In 1940, American physicists Edwin McMillan and Philip Abelson, at the University of California, Berkeley, did succeed in creating element 93, in the same way Fermi had tried. Since uranium was named after the planet Uranus, McMillan and Abelson named the new element **neptunium**, after the planet next furthest from the Sun, Neptune.

The next element, **plutonium** (element 94), was created later in 1940 by a team that included McMillan and was headed by American nuclear chemist Glenn Seaborg, also at Berkeley. Plutonium was first discovered after it was produced as the result of the beta decay of the newly created neptunium; in the same year, the team also produced plutonium by bombarding uranium with deuterons (a deuteron is a particle composed of one proton and one neutron). The new element was named after Pluto, which was then considered to be the planet beyond Neptune (today, it is classed as a dwarf planet). The longest-lived plutonium isotope has a half-life of more than 80 million years – surprisingly long, perhaps, but short enough that any plutonium present when Earth was formed has long ago decayed.

Physicists quickly realized that the isotope plutonium-239 could undergo fission in much the same way as uranium does, and it could be made in quantities large enough to make a bomb. Plutonium can be produced in a uranium-powered nuclear reactor, and fissionable plutonium-239 was used in the first test of an atomic bomb, in July 1945. The bomb dropped on Nagasaki, Japan, little more than a month later, was a plutonium bomb of the same design (the first bomb, dropped on another Japanese city, Hiroshima, relied on the fission of uranium, not plutonium). Most nuclear weapons made to date have plutonium-239 as the fissile material. A range of plutonium isotopes are created inside nuclear reactors, whose fuel is uranium-235. Most fuels are reprocessed once the uranium is spent, to recover the plutonium – in particular plutonium-239 – which is then mixed in with fresh uranium fuel. Another plutonium isotope, plutonium-238, is commonly used in power supplies aboard space probes, in which the energy released by the radioactivity produces electrical power.

---

93 ☢

Np

Neptunium

ATOMIC NUMBER: 93
ATOMIC WEIGHT: (237)
DISCOVERY: 1940
HALF-LIFE OF
LONGEST-LIVED
ISOTOPE: 2.1 million
years

---

94 ☢

Pu

Plutonium

ATOMIC NUMBER: 94
ATOMIC WEIGHT: (244)
DISCOVERY: 1940
HALF-LIFE OF
LONGEST-LIVED
ISOTOPE: 80.8 million
years

Glenn Seaborg's team was also responsible for the next two transuranium elements: **americium** (element 95) and **curium** (element 96). Both elements were created by bombarding plutonium-239 – not with neutrons, but with alpha particles. (An alpha particle is a clump of two protons and two neutrons ejected by unstable nuclei during a process called alpha decay.) Alpha particles were accelerated to high energy inside a device called a cyclotron, and they emerged at high speed to strike a platinum plate that was coated with plutonium nitrate. The element curium was made first, in July 1944, and was named after Marie and Pierre Curie (see page 37). Three months later, americium was produced and identified, and was named after the Americas. Smoke detectors contain a tiny amount (less than one-millionth of a gram) of americium-241, which undergoes alpha decay. The resulting alpha particles ionize the air inside the device, allowing a tiny electric current to flow through the air; when smoke particles enter, they absorb some alpha particles, interrupting the current and triggering the alarm.

The next two elements, **berkelium** (element 97) and **californium** (element 98) were both produced in late 1949 to early 1950 – again by a team headed by Glenn Seaborg at the University of California, Berkeley. To make element 97, the team bombarded americium (element 95) with alpha particles accelerated in their cyclotron. The element's name is derived from Berkeley, the university's home city. Berkelium does not have any applications outside scientific research.

Element 98, californium – named after the state of California – was made in the same way, but by using curium (element 96) as the target. Most californium isotopes can decay either by alpha decay or by spontaneous fission, in which the large nucleus splits into two smaller fragments and releases several neutrons. As a result, even a tiny sample of californium produces a consistent supply of neutrons. The most prodigious neutron-emitting californium isotope is californium-252, which is used in treating various cancers, in metal detectors that can find buried land mines, in a kind of "neutron X-ray" device that can highlight dangerous imperfections in aircraft parts, and as a "starter" for nuclear reactors.

The elements **einsteinium** (element 99) and **fermium** (element 100) were both first detected in the fallout of the world's first ever test of a hydrogen bomb, in 1952. In a hydrogen bomb, hydrogen fuses to make helium and releases huge amounts of energy – but the reaction can only start with extreme heat and pressure, which is achieved by a small uranium or plutonium fission bomb. The uranium was the starting point for the new elements.

A team at Berkeley, this time headed by American physicist Albert Ghiorso, found the new elements and worked out that they had been produced after uranium nuclei had absorbed several neutrons, the resulting nucleus undergoing several successive beta decays, adding one to its atomic number each time. In 1954, Ghiorso's team managed to create both the new elements, by bombarding uranium with ions of nitrogen accelerated in the same cyclotron that was used to create elements 95 to 98. The elements were named after the German-born physicist Albert Einstein and Italian-born physicist Enrico Fermi. Ghiorso and Seaborg were also members of the team that created the next element, **mendelevium** (element 101), in 1955. Again, they used the cyclotron at the University of California, Berkeley, this time firing the alpha particles at a target of einsteinium.

The elements from 102 upwards have all been synthesized by bombarding heavy elements, including lead, bismuth and plutonium, with fairly heavy ions, including ions of

95 Am
Americium

ATOMIC NUMBER: 95
ATOMIC WEIGHT: (243)
DISCOVERY: 1944
HALF-LIFE OF LONGEST-LIVED ISOTOPE: 7,370 years

96 Cm
Curium

ATOMIC NUMBER: 96
ATOMIC WEIGHT: (247)
DISCOVERY: 1944
HALF-LIFE OF LONGEST-LIVED ISOTOPE: 15.6 million years

97 Bk
Berkelium

ATOMIC NUMBER: 97
ATOMIC WEIGHT: (247)
DISCOVERY: 1949
HALF-LIFE OF LONGEST-LIVED ISOTOPE: 1,380 years

98 Cf
Californium

ATOMIC NUMBER: 98
ATOMIC WEIGHT: (251)
DISCOVERY: 1950
HALF-LIFE OF LONGEST-LIVED ISOTOPE: 898 years

99 Es
Einsteinium

ATOMIC NUMBER: 99
ATOMIC WEIGHT: (252)
DISCOVERY: 1952
HALF-LIFE OF LONGEST-LIVED ISOTOPE: 471.7 days

100 Fm
Fermium

ATOMIC NUMBER: 100
ATOMIC WEIGHT: (257)
DISCOVERY: 1952
HALF-LIFE OF LONGEST-LIVED ISOTOPE: 100.5 days

101 Md
Mendelevium

ATOMIC NUMBER: 101
ATOMIC WEIGHT: (258)
DISCOVERY: 1955
HALF-LIFE OF LONGEST-LIVED ISOTOPE: 51.5 days

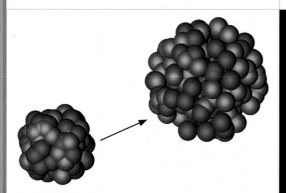

Left: A fast-moving nucleus of an element, such as calcium, bombards a heavier nucleus, such as bismuth.

Right: The 60-inch cyclotron at the University of California, Berkeley, which was used to discover several transuranium elements.

102 ☢	103 ☢
No	**Lr**
Nobelium	**Lawrencium**
ATOMIC NUMBER: 102	ATOMIC NUMBER: 103
ATOMIC WEIGHT: (259)	ATOMIC WEIGHT: (262)
DISCOVERY: 1956	DISCOVERY: 1961
HALF-LIFE OF LONGEST-LIVED ISOTOPE: 58 minutes	HALF-LIFE OF LONGEST-LIVED ISOTOPE: 1 hour 35 minutes

104 ☢	105 ☢
Rf	**Db**
Rutherfordium	**Dubnium**
ATOMIC NUMBER: 104	ATOMIC NUMBER: 105
ATOMIC WEIGHT: (267)	ATOMIC WEIGHT: (268)
DISCOVERY: 1969 – possibly 1966	DISCOVERY: 1960s – possibly 1967
HALF-LIFE OF LONGEST-LIVED ISOTOPE: 1 hour 20 minutes	HALF-LIFE OF LONGEST-LIVED ISOTOPE: 29 hours

106 ☢	107 ☢
Sg	**Bh**
Seaborgium	**Bohrium**
ATOMIC NUMBER: 106	ATOMIC NUMBER: 107
ATOMIC WEIGHT: (269)	ATOMIC WEIGHT: (270)
DISCOVERY: 1974	DISCOVERY: 1981 – and possibly earlier
HALF-LIFE OF LONGEST-LIVED ISOTOPE: 2 minutes 5 seconds	HALF-LIFE OF LONGEST-LIVED ISOTOPE: 2.1 million years.

calcium, nickel and even lead. In many cases, the discoveries were based on the production of just a few atoms, or even just a single one.

In 1958, Albert Ghiorso and Glenn Seaborg used a new apparatus, the heavy ion linear accelerator (HILAC), to bombard curium with ions of the element carbon. The result was element 102, now called **nobelium**. But another team had already created the element two years earlier, at the Joint Institute for Nuclear Research in Dubna, Russia (then in the Soviet Union). The Russian team, headed by physicist Georgy Flyorov (sometimes written Flerov), had produced the new element by bombarding plutonium with ions of oxygen. The previous year, in 1957, a team at the Nobel Institute of Physics, in Stockholm, Sweden, claimed that they had produced element 102 – and they suggested the name nobelium, after Alfred Nobel (see page 116). Although their claim is now known to have been false, the name still stands.

Element 103, **lawrencium**, was first produced in 1961, by a team led by Albert Ghiorso at Berkeley, by bombarding californium with ions of the element boron. It is named after the inventor of the cyclotron, American physicist Ernest Lawrence.

Elements 104, 105 and 106 were the subject of considerable controversy over discovery and naming. In each case, the Dubna team, headed by Georgy Flyorov, reported the discovery first – in 1964, 1967 and 1974 respectively – but their results were disputed by Albert Ghiorso's team at Berkeley. All these elements have since been produced and studied in more detail. The International Union of Pure and Applied Chemistry (IUPAC) decided upon names for these elements as recently as 1997. Element 104 is **rutherfordium**, after New Zealand-born physicist Ernest Rutherford, who discovered the atomic nucleus. Element 105 is **dubnium**, after Dubna, the town where Flyorov's team worked. Element 106 is named **seaborgium**, after Glenn Seaborg; this is the only time an element has been named after someone who was alive at the time (Seaborg died in 1999).

There was controversy, too, over the discovery of **bohrium** (element 107). A team led by German physicist Peter Armbruster, at the Gesellschaft für Schwerionenforschung (GSI, the Society for Heavy Ion Research) in Darmstadt, Germany, and the team in Dubna, Russia, both made claims to it in the late 1970s. The first convincing synthesis of element 107 was carried out by the GSI team, in 1981, and credit is normally given to them. The element is named after Danish physicist Niels Bohr (see page 17).

Armbruster's team at the GSI were also first to produce element 108, **hassium**, named after the Latin name for the German state Hesse, where the laboratory is located. The same is true of **meitnerium** (element 109), which is named after German physicist Lise Meitner (see page 96).

Various isotopes of element 110, **darmstadtium**, were produced at both the GSI and Dubna from 1987 onwards; the element's name is derived from the city of Darmstadt, where the GSI is located. Element 111, **roentgenium**, was first produced at the GSI in 1994. It is named after German physicist Wilhelm Conrad Röntgen, known for his pioneering work with X-rays. Element 112, **copernicium** – also first created at GSI in 1996 – is named after the sixteenth-century Polish astronomer Nicolaus Copernicus.

The remaining elements so far discovered (up to 118) were mostly created either in Dubna or at the Lawrence Livermore National Laboratory in California, some at GSI and one at the Advanced Science Institute at Rikagaku Kenkyujo (RIKEN), near Tokyo, Japan.

In 2004, IUPAC brought in a system of temporary names for new elements, based on a mixture of Greek and Latin numbering. So, element 113 is currently **ununtrium** (1-1-3-ium), 115 is **ununpentium**, 117 is **ununseptium** and 118 is **ununoctium**. Once an element's existence is officially confirmed, a permanent name can be assigned – but it has to be agreed by IUPAC and the International Union of Pure and Applied Physics (IUPAP). So, for example, in May 2012, element 114 became **flerovium**, after Georgy Flyorov, who died in 1990), and element 116 became **livermorium**, after the Lawrence Livermore Laboratory, which was involved in its discovery.

Efforts are still continuing to produce new, heavier elements. If such elements can be synthesized, they will almost certainly be extremely unstable, and would exist only very fleetingly – although there is a possibility that "islands of stability" exist, in which case scientists may even create fairly stable superheavy atoms. For example, some theoretical models of the nucleus suggest that certain isotopes of the proposed element 126 (unbihexium) might be relatively stable.

Left: American nuclear chemist Glenn Seaborg, who was involved in the discoveries of several transuranium elements. Seaborg was the only person to have an element (seaborgium) named after him while still alive.

108 ☢ **Hs** Hassium	109 ☢ **Mt** Meitnerium
ATOMIC NUMBER: 108 ATOMIC WEIGHT: (269) DISCOVERY: 1984 HALF-LIFE OF LONGEST-LIVED ISOTOPE: 11 minutes (approximate)	ATOMIC NUMBER: 109 ATOMIC WEIGHT: (278) DISCOVERY: 1982 HALF-LIFE OF LONGEST-LIVED ISOTOPE: 8 seconds (approximate)

110 ☢ **Ds** Darmstadtium	111 ☢ **Rg** Roentgenium	112 ☢ **Cn** Copernicium
ATOMIC NUMBER: 110 ATOMIC WEIGHT: (281) DISCOVERY: 1994 HALF-LIFE OF LONGEST-LIVED ISOTOPE: Possibly around 4 minutes	ATOMIC NUMBER: 111 ATOMIC WEIGHT: (281) DISCOVERY: 1994 HALF-LIFE OF LONGEST-LIVED ISOTOPE: 26 seconds (approximate)	ATOMIC NUMBER: 112 ATOMIC WEIGHT: (285) DISCOVERY: 1996 HALF-LIFE OF LONGEST-LIVED ISOTOPE: Possibly around 9 minutes

113 ☢ **Uut** Ununtrium	114 ☢ **Fl** Flerovium	115 ☢ **Uup** Ununpentium
ATOMIC NUMBER: 113 ATOMIC WEIGHT: (286) DISCOVERY: 2003 HALF-LIFE OF LONGEST-LIVED ISOTOPE: 20 seconds (approximate)	ATOMIC NUMBER: 114 ATOMIC WEIGHT: (289) DISCOVERY: 1998/99 HALF-LIFE OF LONGEST-LIVED ISOTOPE: Possibly around 1 minute	ATOMIC NUMBER: 115 ATOMIC WEIGHT: (289) DISCOVERY: 2003 HALF-LIFE OF LONGEST-LIVED ISOTOPE: 0.2 seconds (approximate)

116 ☢ **Lv** Livermorium	117 ☢ **Uus** Ununseptium	118 ☢ **Uuo** Ununoctium
ATOMIC NUMBER: 116 ATOMIC WEIGHT: (293) DISCOVERY: 2000 HALF-LIFE OF LONGEST-LIVED ISOTOPE: 60 milliseconds (approximate)	ATOMIC NUMBER: 117 ATOMIC WEIGHT: (294) DISCOVERY: 2010 HALF-LIFE OF LONGEST-LIVED ISOTOPE: 78 milliseconds (approximate)	ATOMIC NUMBER: 118 ATOMIC WEIGHT: (294) DISCOVERY: 2002 HALF-LIFE OF LONGEST-LIVED ISOTOPE: 1 millisecond (approximate)

Index

Page numbers in **bold** type refer to main entries

Credits

Every effort has been made to acknowledge correctly and contact the source and/or copyright holder of each picture, and Carlton Books Limited apologizes for any unintentional errors or omissions, which will be corrected in future editions of this book.

The majority of photographs were supplied by Science Photo Library with the exception of the page 44. (left) Intel, 54. (left) Getty Images/Joe Raedle, 105. (bottom right) Thinkstock, 107 (bottom right) Thinkstock/Comstock, 108. Thinkstock/Digital Vision, 133. (bottom right) Getty Images/Natasja Weitsz, 137. (bottom) Gore-Tex

Science Photo Library contributor acknowledgements: 14. Shelia Terry, 17. (top & bottom) Ria Novosti, 18. Adam Block, 19. (top) Tony Craddock, (bottom left) Martyn F. Chillmaid, (bottom) Pasieka, 20. (top left) Kenneth Eward/Biografx, 20. (top left) US Navy, (bottom) AJ Photo, 21. (top) Martin Bond, (bottom) US Department of Energy, 23. (bottom left) Eye of Science, 24. (left) David Taylor, 25. (top right) Bill Beatty, Visuals Unlimited, 25. (top left) David Nunek, Alexis Rosenfeld (bottom), 26. (right) Andrew Lambert Photography, (bottom) Charles D. Winters, 27. (top right) David Cattlin, (bottom left) Cristina Pedrazzini, Sciepro (bottom right), 28. (right) E.R. Degginer, 29. (top right) Andrew Brookes, National Physical Laboratory, 31. (top) Russ Lappa, (centre) NASA, Roberto De Gugliemo (bottom), 32. (left) Jerry Mason, (right) Russ Lappa, 33. (top) Herve Conge, ISM (centre) Charles D. Winters, 34. (top) Steve Allen, (bottom left) Steve Gschmeissner, (bottom right) Scott Camazine, 35. (bottom) Tony Craddock, 36. Sovereign, ISM, 37. J.C. Revy, ISM (top), (bottom left) C.Powell, P.Fowler & D.Perkins, (bottom right) Health Protection Agency, 41. (left) Custom Medical Stock Photo, 42. (bottom right) Klaus Guldbrandsen, 42. (top) Thedore Gray, Visuals Unlimited, (left) Carlos Dominguez, 43. (left) Joel Arem, 45. (bottom) Charles D. Winters, 46. (top) Andrew Lambert Photography, (right) Andrew Lambert Photography, (bottom left) William Lingwood, 47. Wim Van Cappellen/Reporters, 48. (left) David Parker, (right) NASA, 49. (top) Giphotostock, (bottom left) Russell Lappa, (bottom right) James King-Holmes, 50. (left) Charles D. Winters, 51. (centre) Joel Arem, (bottom) Tom Burnside, (top) Mark Williamson, 52. (left) Shelia Terry, 54. (right) D.Roberts, 55. (bottom) Charles D. Winters, 56. (top) Science Vu, Visuals Unlimited, (bottom) ISM, 57. (top) Science Source, 58. (left) US Geological Survey, (bottom right) Alex Bartel, 59. (top) Charles D. Winters, (centre) Ferrofluid, (bottom) Theodore Gray, 60. (top) Steve Gschmeissner, (bottom) David Taylor, 61. (top) David R. Frazier, 62. (top) Philippe Psaila, 63. (top) Dirk Wiersma, (top) Sovereign/ISM, 64. (bottom) Gregory Davies, Medinet Photographics, 65. Dr Tim Evans (top), 66. (bottom) Martin Bond, 67. (top) Charles D. Winters (bottom), Stefan Diller, 68. (bottom left) Mark Sykes, (bottom right) Martyn F. Chillmaid, (top) Ria Novosti, 69. (bottom) Manfred Kage, 70. (top) Ken Lucas, Visuals Unlimited, (bottom) Malcolm Fielding, Johnson Matthey PLC, 71. (top right) Jim Amos, (centre) Dr. P. Marazzi, (bottom) Jim Amos, 72. (left) R. Maisonneuve, Publiphoto Diffusion ,(right) Dirk Wiersma, 73. (top) David Parker, (bottom) Simon Lewis, 74. (left) David R. Frazier, (right) Theodore Gray, Visuals Unlimited, 75. (top) Tek Image, (bottom) Dr P. Marazzi, 76. (left) Photo Researchers, 77. (top) Patrick Landman, (centre) Eye of Science, (bottom) Maximilian Stock Ltd, 78. (bottom) Herman Eisenbeiss, 79. (top) Peidong Yang/UC Berkeley, (bottom) Francoise Sauze, 80. (centre) Rich Treptow, (top) National Institute of Standards and Technology (bottom), Max-Planck-Institute for Metallurgy, 81. (top) Robert Brook, (centre) Cordelia Molloy, 82. (left) Charles D. Winters, (right) Charles D. Winters, 83. (top) Gustoimages, (bottom) CNRI, 87. (bottom left) Gustoimages, (bottom right) Eye of Science, 90. (centre) Sovereign, ISM, 93. (top) Andrew Brookes, National Physical Laboratory, 96. (left) Patrick Landmann, (right) Dirk Wiersma, 97. (bottom left) Alexis Rosenfeld, 99. (left) Dirk Wiersma, 101. (top) Martyn F. Chillmaid, 102. (bottom) Drs A. Yazdani & DJ Hornbaker, (top) Kenneth Eward, 101. (bottom left) Natural History Museum, (bottom right) Mark Williamson, 105. (top) Raul Gonzalez Perez, (bottom left) Carol & Mike Werner, Visuals Unlimited, 106. (top left) James Holmes/Zedcor, (top right) J.Bernholc Et Al, North Caroline State University, 107. (top left) Cordelia Molloy, (bottom left) Dr Tim Evans, 108. (bottom left) Richard Folwell, 110. (top) Manfred Kage, (bottom) Klaus Guldbrandsen, 112. (top) Johnny Grieg, (centre) Erich Schrempp, 113. (top) Patrick Landmann, (bottom) Martyn F. Chillmaid, 115. (bottom left) Charles D. Winters, (right) Martyn F. Chillmaid,116. (top) Gordon Garradd, (bottom) Wally Eberhart, Visuals Unlimited, 117. (top) Ian Gowland, (bottom) Thedore Gray, Visuals Unlimited, 118. (bottom) Laguna Design, 119. (top) Charles D. Winters, (bottom) Dirk Wiersma, 120. (top) Zephyr, (bottom) Dirk Wiersma, 123. (top) Dr. Tim Evans, (centre) Martyn F. Chillmaid, (bottom) David Woodfall Images, 124. (top) Dirk Wiersma, (bottom) Richard Bizley, 125. (top left) NASA, (top right) Marti Miller, Visuals Unlimited, (bottom right) British Antarctic Survey, 126. (top) Ton Kinsbergen, (bottom) Charles D. Winters, 127. (left) Charles D. Winters, (right) Bernhard Edmaier, 128. (top) Mark Sykes, (bottom left) Simon Fraser, (bottom centre) Bernhard Edmaier, 129. (top) NASA, (bottom) Monty Rakusen, 130. (bottom) Chris Knapton, 131. (top) Alan & Linda Detrick, (bottom) C.S. Langlois, Publiphoto Diffusion, 132. (top) Dirk Wiersma, 133. (top) Astrid & Hanns-Frieder Michler, (bottom left) Rich Treptow, 135. Dirk Wiersma, 136. (top) Charles D. Winters, 137. (top) Ria Novosti, 138. (left) Charles D. Winters, (bottom right) CC Studio, 139. (top) Steve Horrell, 140. (bottom) Laguna Design, (top left) Victor de Schwanberg, (top right) Romilly Lockyer, Cultura, (bottom) 141. Charles D. Winters, 142. (bottom) Adrian Bicker, (right) Charles D. Winters, 143. Dr. Ken Greer, Visuals Unlimited, 145. (top) Thedore Gray, Visuals Unlimited, (bottom) European Space Agency, 146. (bottom left) Philippe Psaila, (bottom right) Alexander Tsiaras, 147. (bottom) Thedore Gray, Visuals Unlimited, 148. (bottom) Thedore Gray, Visuals Unlimited, (top) Peter Menzel, 149. John Reader, 150. (top) Peter Menzel, (bottom) Thedore Gray, Visuals Unlimited, 151. (top) Theodore Gray, Visuals Unlimited, 151. (bottom) NASA/JPL, 152. National Cancer Institute, 154. Los Alamos National Laboratory, 156 & 157. Lawrence Berkeley Laboratory

Special thanks to Anna Bond at Science Photo Library for her hard work on this book. All other photographs/illustrations courtesy of the author/Carlton Publishing Group.

Publishing Credits

Executive Editor: **Gemma Maclagan Ram**
Copyeditor: **Katie John**
Proof reader: **Martyn Page**
Design Direction: **Darren Jordan**
Design: **Harj Ghundale**
Cover Design: **Grade**
Picture Research Manager: **Steve Behan**
Production Manager: **Maria Petalidou**